THE
ELEMENTS OF AEROFOIL
AND AIRSCREW THEORY

THE
ELEMENTS OF AEROFOIL
AND AIRSCREW THEORY

by

H. GLAUERT

Second Edition

CAMBRIDGE
UNIVERSITY PRESS

CAMBRIDGE UNIVERSITY PRESS
Cambridge, New York, Melbourne, Madrid, Cape Town, Singapore,
São Paulo, Delhi, Dubai, Tokyo, Mexico City

Cambridge University Press
The Edinburgh Building, Cambridge CB2 8RU, UK

Published in the United States of America by
Cambridge University Press, New York

www.cambridge.org
Information on this title: www.cambridge.org/9780521274944

First published 1926
Reprinted 1930, 1937, 1942
Second edition 1947
Reprinted 1948, 1959
Reissued 1983 in the Cambridge Science Classics Series
Reprinted 1986, 1993

A catalogue record for this publication is available from the British Library

Library of Congress catalogue card number: 82–22233

ISBN 978-0-521-27494-x Hardback
ISBN 978-0-521-27494-4 Paperback

PREFACE

The aim of aerofoil theory is to explain and to predict the force experienced by an aerofoil, and a satisfactory theory has been developed in recent years for the lift force in the ordinary working range below the critical angle and for that part of the drag force which is independent of the viscosity of the air. Considerable insight has also been obtained into the nature of the viscous drag and into the behaviour of an aerofoil at and above the critical angle, but the theory remains at present in an incomplete state. The problem of the airscrew is essentially a part of aerofoil theory, since the blades of an airscrew are aerofoils which describe helical paths, and a satisfactory theory of the propulsive airscrew has been developed by extending the fundamental principles of aerofoil theory.

The object of this book is to give an account of aerofoil and airscrew theory in a form suitable for students who do not possess a previous knowledge of hydrodynamics. The first five chapters give a brief introduction to those aspects of hydrodynamics which are required for the development of aerofoil theory. The following chapters deal successively with the lift of an aerofoil in two dimensional motion, with the effect of viscosity and its bearing on aerofoil theory, and with the theory of aerofoils of finite span. The last three chapters are devoted to the development of airscrew theory.

In accordance with the object of the book, complex mathematical analysis has been avoided as far as possible and in a few cases results have been quoted without proof, the reader being referred for further details to standard text-books or to original papers on the subject.

My thanks are due to my wife for her assistance in preparing a number of the figures and in reading the proof sheets, and to the Cambridge University Press for their care and vigilance in passing the book through the proof stage.

H. G.

Farnborough, April 1926.

PREFACE TO SECOND EDITION

Great advances in the theory of aeronautics have taken place since the first edition of this book by my late husband appeared in 1926, but the more fundamental parts of the theory, which are the subject of this book, remain in large measure unchanged. Particularly important advances have been made in the theory of viscous motion and of the flow in the boundary layer. At my request Mr H. B. Squire of the Royal Aircraft Establishment, Farnborough, who was a colleague of my husband, has prepared a set of notes which appear as an Appendix to the present edition and these notes indicate where important developments have taken place and where further information on the subject matter can be found. I am most grateful to Mr Squire for his assistance and desire to tender him my sincere thanks.

In preparing this second edition the opportunity has been taken to replace the non-dimensional k coefficients by the now more generally accepted C coefficients and my son, M. B. Glauert, has undertaken the necessary revision. One or two other minor changes have been made and a bibliography of some of the more important modern books on aerodynamics has been added.

M. G.

Cambridge 1946

CONTENTS

REFERENCES

The following abbreviations are used:

RM   = *Reports and Memoranda of the Aeronautical Research Committee.*

NACA   = *Reports of the National Advisory Committee for Aeronautics (U.S.A.).*

ZFM   = *Zeitschrift für Flugtechnik und Motorluftschiffahrt.*

ZAMM = *Zeitschrift für angewandte Mathematik und Mechanik.*

FD   = *Modern Developments in Fluid Dynamics.*

CHAPTER I

INTRODUCTION

1·1. It is a fact of common experience that a body in motion through a fluid experiences a resultant force which, in most cases, is mainly a resistance to the motion. A class of body exists, however, for which the component of the resultant force normal to the direction of motion is many times greater than the component resisting the motion, and the possibility of the flight of an aeroplane depends on the use of a body of this class for the wing structure.

A wing or aerofoil has a plane of symmetry passing through the mid-point of its span, and the direction of motion and the line of action of the resultant force usually lie in this plane. The section of an aerofoil by a plane parallel to the plane of symmetry is of an elongated shape with a rounded leading edge and a fairly sharp trailing edge. The chord line of an aerofoil is defined as the line joining the centres of curvature of the leading and trailing edges and the projection of the aerofoil section on this line is defined as the chord length. Aerofoil sections which are used on airscrews are flat over most of the lower surface and the chord line of these sections is usually taken along the flat under-surface of the

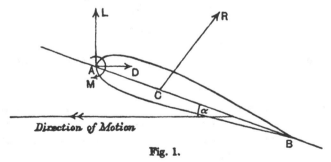

Fig. 1.

aerofoil. The angle of incidence α of an aerofoil is defined as the angle between the chord and the direction of motion relative to the fluid, and the centre of pressure C of an aerofoil is defined as the point in which the line of action of the resultant force R intersects the chord AB (fig. 1). The resultant force is resolved into two components, the lift L at right angles to the direction of motion and the drag D parallel to that direction but opposing the motion. It is customary to use the leading edge A of the chord as a point of reference and the resultant force has a moment M about this point, whose sense is such that a positive moment tends to increase the angle of incidence*. The magnitude of this moment is

$$M = -AC \, (L \cos \alpha + D \sin \alpha),$$

where AC is the distance of the centre of pressure behind the leading edge of the chord.

The resultant force on an aerofoil of a given shape at a definite angle of incidence depends mainly on the density ρ of the fluid, the relative velocity V of the aerofoil and the fluid, and some typical length l of the aerofoil. These three quantities can be combined in the unique form $l^2 \rho V^2$ to give the dimensions of a force, and non-dimensional coefficients of lift and drag may be defined by dividing the force components by this product. The standard lift and drag coefficients of an aerofoil are defined by the equations

$$L = C_L \cdot \tfrac{1}{2}\rho V^2 S,$$
$$D = C_D \cdot \tfrac{1}{2}\rho V^2 S,$$

where S is the maximum projected area of the aerofoil which, in the case of a rectangular aerofoil, is the product of the chord and the span. The corresponding definition for the moment coefficient is

$$M = C_M \cdot \tfrac{1}{2}\rho V^2 S c,$$

where c is the chord of the aerofoil. These definitions are not unique and the older British practice is to use ρV^2 instead of the dynamic pressure $\tfrac{1}{2}\rho V^2$. This gives coefficients k_L, k_D and k_m half as large as those above.

The lift and drag coefficients of an aerofoil are functions of the angle of incidence and fig. 2 shows the curves for a typical aerofoil, the drag being drawn to five times the scale

* See Note 1 of Appendix.

of the lift. The lift coefficient varies linearly with the angle
of incidence for a certain range and then attains a maximum

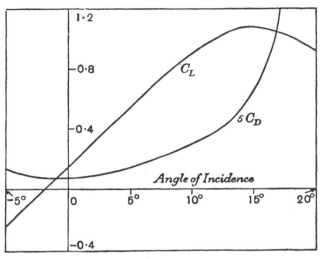

Fig. 2.

value at the critical angle of incidence. The important working range of an aerofoil is represented by the linear part of
the lift curve and in this range the drag is small compared
with the lift, but on approaching the critical angle the drag
increases rapidly.

Fig. 3 shows the variation of the position of the centre of
pressure, the distance of the centre of pressure behind the
leading edge of the aerofoil being expressed as a fraction of
the chord. Analytically this centre of pressure coefficient is

$$\frac{AC}{AB} = -\frac{C_M}{C_L \cos \alpha + C_D \sin \alpha} = -\frac{C_M}{C_L} \text{(approximately)},$$

and theory and experiment agree in showing that the moment
coefficient varies in a linear manner with the lift coefficient
below the critical angle. The centre of pressure of an aerofoil
section normally moves backwards as the angle of incidence
decreases and tends to infinity at the negative angle of incidence for which $(C_L \cos \alpha + C_D \sin \alpha)$ vanishes, i.e. when

the resultant force on the aerofoil is parallel to the chord. This angle of incidence is approximately equal to the angle at which the lift vanishes.

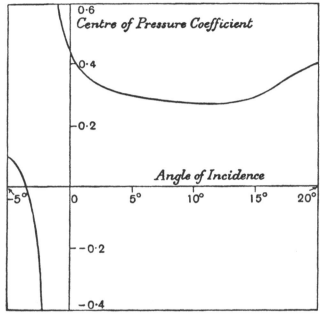

Fig. 3.

The main object of aerofoil theory is to explain and to predict the lift and drag experienced by an aerofoil, and a satisfactory theory has been developed in recent years for the ordinary working range below the critical angle. The determination of the maximum lift of an aerofoil and of the critical angle at which it occurs is not yet possible, although some insight has been obtained into the cause of the phenomenon.

1·2. *The development of aerofoil theory.*

The explanation of the lift force of an aerofoil depends essentially on the nature of the fluid, and the difficulty of obtaining a satisfactory theory is associated with the difficulty of defining the essential characteristics of the fluid in a simple and reliable manner.

An early attempt to develop a theory of the force on an inclined flat plate is due to Newton, who assumed that the fluid consisted of a large number of solid corpuscles. These corpuscles were assumed to be inelastic and, on striking the plate, the component of their velocity normal to the plate would be destroyed. The mass of fluid meeting a plate of area S at an angle of incidence α in unit time is $S\rho V \sin \alpha$ and the velocity normal to the plate is $V \sin \alpha$. Hence the plate would experience a force normal to its surface of magnitude

$$R = S\rho V^2 \sin^2 \alpha.$$

If the corpuscles are assumed to be perfectly elastic, this force is doubled, but in either case the force given by this theory at small angles of incidence is too small. The estimate of the drag of a flat plate set normal to the direction of motion is more satisfactory and is of the correct order of magnitude.

A better definition of the characteristics of a fluid was obtained by regarding the fluid as a continuous homogeneous medium. An essential characteristic of a fluid is that it cannot support tangential stresses in a state of equilibrium, but when adjacent layers of the fluid are in relative motion tangential stresses exist and oppose the motion. This characteristic is due to the internal friction or viscosity of the fluid. The viscosity of the air is small and may be neglected in a large number of problems, but at times the viscosity is of fundamental importance and in all cases it appears to exert a determining influence on the type of motion which occurs, even when the motion proceeds exactly as in a non-viscous fluid. Another characteristic of a fluid is its compressibility, which is negligible for a liquid but important for a gas. The density of the air must be regarded in general as a function of the pressure and temperature, but the variations of the pressure in the flow past a body are sufficiently small to justify the assumption that the density of the air is constant. This assumption, however, ceases to be valid when the velocity of the flow becomes comparable in magnitude with the velocity of sound and allowance must then be made for the compressibility of the air.

These considerations led to the conception of the air as a perfect fluid, i.e. as a continuous incompressible non-viscous medium. The development of the theory of fluid motion has been based on this conception and the results deduced from the theory are of great value in many cases. Unfortunately the theory led to the astonishing conclusion that a body in motion through a perfect fluid does not experience any resultant force.

An attempt to surmount this discrepancy between theory and fact was made by Helmholtz and Kirchhoff by assuming that the flow past a body, instead of passing round the whole surface, leaves a wake or dead-water region behind the body. This method of discontinuous flow* has been applied to an inclined flat plate in two dimensional motion, which is equivalent to an aerofoil of infinite span, and gives a resultant force normal to the surface of magnitude

$$R = \frac{\pi \sin \alpha}{4 + \pi \sin \alpha} S \rho V^2.$$

This force is of the correct order of magnitude for small angles of incidence and also for a flat plate set normal to the direction of motion, but the actual numerical values are not in good agreement with experimental results.

A lift force can also be obtained in a perfect fluid if the flow is assumed to have a tendency to circulate round the body, and modern aerofoil and airscrew theory is based on this conception. The development of the theory for an aerofoil of infinite span, which corresponds to motion in two dimensions, is due in the first place to Kutta† and Joukowski‡, and

---

* For the development of the theory see Lamb, *Hydrodynamics*, § 73 *et seq.*

† "Auftriebskräfte in strömenden Flüssigkeiten," *Illust. aeronaut. Mitteilungen*, 1902. "Über eine mit den Grundlagen des Flugproblems in Beziehung stehende zwei dimensionale Strömung," *Sitzb. d. k. Bayr. Akad. d. Wiss.* 1910.

‡ "Über die Konturen der Tragflächen der Drachenflieger," *ZFM*, 1910.

the extension to the general case in three dimensions, which follows the general lines suggested by Lanchester*, is due to Prandtl†. The theory gives results in close agreement with experiment but there remains the difficulty of explaining the origin of the circulation. In a perfect fluid this circulation could not develop and it must be ascribed to the action of the viscosity in the initial stages of the motion.

The general aerofoil theory indicates that there is a drag force (induced drag) associated with the lift of an aerofoil, but for motion in two dimensions this induced drag becomes zero and it is again necessary to turn to the viscosity of the fluid for the explanation of the small drag force (profile drag) which actually exists. The development of the theory of an aerofoil is therefore based in the first place on the assumption that the air is a perfect fluid, and the viscosity is introduced at a later stage to explain the origin of the circulation and the existence of the profile drag.

1·3. *Atmospheric relationships.*

Although the compressibility of the air can be neglected in most problems of the flow past a body, the density of the air cannot be regarded as an absolute constant but must be determined as a function of the pressure and temperature of the undisturbed air according to the physical law

$$\frac{p}{p_0} = \frac{\rho}{\rho_0} \frac{\theta}{\theta_0},$$

where p is the pressure, ρ the density and θ the absolute temperature.

In the atmosphere the pressure and density are connected with the height above the ground by the equation

$$\frac{dp}{dh} = -g\rho,$$

but to determine the conditions at any height it is necessary to know also the relationship between the temperature and

---

* *Aerodynamics*, 1907. An account of his theory in a less developed form was given by Lanchester to the Birmingham Natural History and Philosophical Society in 1894.

† "Tragflügeltheorie," *Göttingen Nachrichten*, 1918 and 1919.

the height. This relationship will vary at different places and at different times, but a standard atmosphere has been adopted by many countries as a basis of comparison. The standard atmosphere is defined by a pressure of 760 mm. of mercury (14·7 lb. per sq. in.) at ground level and by the temperature law
$$T = 15 - 0·0065z,$$
where T is the temperature in degrees centigrade and z is the height in metres. This law represents the average conditions in western Europe and is valid up to the height where the temperature ceases to fall on approaching the isothermal layer. The variation of pressure and density with height for the standard atmosphere is given in table 1.

When a change of pressure occurs so rapidly that there is no exchange of heat between adjacent fluid elements, the pressure and density are related by the adiabatic law
$$\frac{p}{p_0} = \left(\frac{\rho}{\rho_0}\right)^\gamma,$$
where γ is the ratio of the two specific heats of the gas and has the numerical value 1·4 for air. The adiabatic law would be satisfied in the atmosphere if the temperature gradient were 3° C. per 1000 ft., and whenever the temperature gradient rises above this value the atmosphere is in an unstable condition which gives rise to convection currents.

Table 1.

Standard Atmosphere.

Height ft.	Pressure p/p_0	Density ρ/ρ_0	Temperature ° C.
0	1·000	1·000	15·0
5,000	0·832	0·862	5·1
10,000	0·688	0·738	− 4·8
15,000	0·565	0·630	− 14·7
20,000	0·460	0·534	− 24·6
25,000	0·372	0·449	− 34·5
30,000	0·298	0·375	− 44·4

1·4. *Units*.

It is customary in aeronautics to express numerical values in British Engineering units and to take the second as the unit of time, the foot as the unit of length and the pound as the unit of force. A new unit of mass becomes necessary, defined by the condition that unit force acting upon unit mass produces unit acceleration. This unit of mass is called the *slug* and is such that a body which weighs W lb. has a mass of W/g slugs ($g = 32 \cdot 2$, approx.).

Continental writers use a similar engineering system in which the second is the unit of time, the metre is the unit of length and the kilogram is the unit of force. The name *newton* has been proposed for the corresponding unit of mass.

The principal relationships between the two systems of units are as follows:

Length	1 m.	$= 3 \cdot 281$ ft.,
Force	1 kg.	$= 2 \cdot 204$ lb.,
Mass	1 newton	$= 0 \cdot 672$ slug,

and the standard density of the air at ground level is 0·00238 slug per cubic foot or 0·125 newton per cubic metre.

CHAPTER II

BERNOULLI'S EQUATION

2·1. *Stream lines and steady motion.*

When a body moves through a fluid with uniform velocity V in a definite direction, the conditions of the flow are exactly the same as if the body were at rest in a uniform stream of velocity V, and it is usually more convenient to consider the problem in the second form. In general therefore the body will be regarded as fixed and the motion of the fluid will be determined relative to the body. A representation of the flow past a body at any instant can be obtained by drawing the stream lines, which are defined by the condition that the direction of a stream line at any point is the direction of motion of the fluid element at that point. In general, the form of the stream lines will vary with the time and so the stream lines are not identical with the paths of the fluid elements. Frequently, however, the flow pattern does not vary with the time and the velocity is constant in magnitude and direction at every point of the fluid. The fluid is then in steady motion past the body and the stream lines coincide with the paths of the fluid elements. The stream lines which pass through the circumference of a small closed curve form a cylindrical surface which is called a stream tube, and since the stream lines represent the direction of motion of the fluid there is no flow across the surface of a stream tube. The theory of the flow past an aerofoil or airscrew is developed almost entirely as a problem of steady motion and, except where otherwise specified, the fluid is regarded as incompressible and non-viscous.

2·2. *Bernoulli's equation.*

In steady motion it is possible to obtain a simple relationship connecting the pressure and velocity at any point of a stream line. The dynamical equation for the motion of a small element of fluid forming part of a stream tube is

$$\rho S v \frac{dv}{ds} = - S \frac{dp}{ds},$$

where S is the cross sectional area of the stream tube at the point under consideration and s is a co-
ordinate measured a-
long the stream tube.
On integrating along
the stream tube

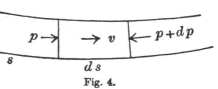

Fig. 4.

$$\tfrac{1}{2}v^2 + \int \frac{dp}{\rho} = \text{constant}$$

in general, and for an incompressible fluid

$$p + \tfrac{1}{2}\rho v^2 = H.$$

This result is known as Bernoulli's equation and the quantity H, which is constant along a stream line, is called Bernoulli's constant or the total pressure head of the fluid. In general, H may have different values for different stream lines, but if the stream lines originate in a region of constant pressure and velocity, it is evident that H will have the same value throughout the fluid. The variation of the value of H for different stream lines, when it occurs, is associated with the presence of vorticity in the fluid (see 4·32), and changes in the value of H may be produced in a real fluid by the action of viscosity.

Bernoulli's equation shows that the pressure of the fluid is greatest where the velocity is least and that H is the maximum pressure which can be attained at any point. This maximum pressure always occurs at some point on the nose of a body where the fluid is brought to rest and the stream divides to pass along the surface of the body. The measurement of the speed of an aircraft depends on this result, since the standard pressure head instrument measures the difference between the total pressure head H and the fluid pressure p. The instrument must be placed with its axis parallel to the direction of the local stream lines and it will then determine the local relative velocity. This velocity may, however, differ from that of the aircraft owing to the disturbance of the general stream caused by the aircraft.

The cross sectional area S of a stream tube is related to the velocity by the condition that $\rho v S$ must be constant, since there is no flow across the surface of the stream tube. Hence for an incompressible fluid S is inversely proportional to v, and the stream tube contracts as the velocity increases. The velocity cannot, however, increase indefinitely since the pressure will become negative when the velocity exceeds the value $\sqrt{2H/\rho}$ and a fluid cannot sustain a negative pressure. To obtain a numerical estimate of this limiting velocity in the air under normal conditions, the value of H may be taken as that of the standard atmospheric pressure (14·7 lb. per sq. in.) and the limiting velocity is then 1340 ft. per sec. This velocity is greater than the velocity of sound and the assumption that the air can be regarded as an incompressible fluid breaks down at a considerably lower velocity.

2·3. *The velocity of sound.*

If a disturbance, such as a sudden increase of pressure, occurs at some point of an incompressible fluid, the disturbance is transmitted instantaneously to all parts of the fluid, but in a compressible fluid the disturbance travels through the fluid in the form of a pressure wave at a definite velocity, which is in fact the velocity of sound in the fluid.

Consider the motion in one dimension along a straight tube of uniform cross sectional area S. If ξ denote the displacement at time t of the particles whose undisturbed position is determined by the coordinate x, then the fluid originally between the limits x and $x + dx$ will at time t lie between the limits

$$x + \xi \quad \text{and} \quad x + \xi + \left(1 + \frac{\partial \xi}{\partial x}\right) dx.$$

The equation of continuity, which expresses the condition that the mass of an element of fluid remains constant, is therefore

$$\rho \left(1 + \frac{\partial \xi}{\partial x}\right) = \rho_0,$$

where ρ_0 is the density of the fluid in the undisturbed state. Now let $\rho = \rho_0(1 + s)$ and then for small disturbances the equation of continuity becomes

$$s = -\frac{\partial \xi}{\partial x}.$$

The dynamical equation for the motion of the fluid along the tube is

$$\rho_0 S\,dx\,\frac{\partial^2 \xi}{\partial t^2} = -S\frac{\partial p}{\partial x}\,dx,$$

or

$$\rho_0\frac{\partial^2 \xi}{\partial t^2} = -\frac{\partial p}{\partial x}.$$

Now the pressure is a function of the density and so for small disturbances

$$p = p_0 + \left(\frac{dp}{d\rho}\right)_0 (\rho - \rho_0)$$

$$= p_0 + s\rho_0 \left(\frac{dp}{d\rho}\right)_0.$$

Hence $$\rho_0\frac{\partial^2 \xi}{\partial t^2} = -\frac{\partial s}{\partial x}\rho_0\left(\frac{dp}{d\rho}\right)_0 = \frac{\partial^2 \xi}{\partial x^2}\rho_0\left(\frac{dp}{d\rho}\right)_0,$$

which may be written as

$$\frac{\partial^2 \xi}{\partial t^2} = c^2\frac{\partial^2 \xi}{\partial x^2}$$

if $$c^2 = \left(\frac{dp}{d\rho}\right)_0.$$

The solution of the differential equation for ξ is

$$\xi = f(x - ct) + F(x + ct),$$

which represents two waves travelling in opposite directions with the velocity c. This velocity c is independent of the type or periodicity of the disturbance and is the velocity of sound in the fluid.

If the temperature of a gas remains constant, the pressure and density are related by Boyle's law

$$\frac{p}{p_0} = \frac{\rho}{\rho_0}.$$

The velocity of sound is then $\sqrt{p_0/\rho_0}$ and for standard atmospheric conditions the numerical value is 945 ft. per sec.

This is considerably below the value determined experimentally and the discrepancy is due to the fact that the temperature does not remain constant during the disturbance. The changes in pressure occur so rapidly that there is no exchange of heat between adjacent fluid elements, and in consequence the pressure and density are related by the adiabatic law

$$\frac{p}{p_0} = \left(\frac{\rho}{\rho_0}\right)^\gamma,$$

where $\gamma = 1\cdot4$ for air. The velocity of sound is therefore $\sqrt{\gamma p_0/\rho_0}$ and the corresponding numerical value is 1120 ft. per sec., which agrees well with the experimental determinations.

In general p_0/ρ_0 is proportional to the absolute temperature θ and the numerical value of c corresponds to the standard ground temperature of 15° C. For any other temperature

$$c = 66\sqrt{\theta} \text{ ft. per sec.,}$$

where θ is the absolute temperature on the centigrade scale.

2·4. *Bernoulli's equation in a compressible fluid.*

The general form of Bernoulli's equation is

$$\tfrac{1}{2}v^2 + \int\frac{dp}{\rho} = \text{constant,}$$

and in a compressible gas the pressure and density are related by the adiabatic law

$$\frac{p}{p_0} = \left(\frac{\rho}{\rho_0}\right)^\gamma.$$

On integrating, Bernoulli's equation becomes therefore

$$\tfrac{1}{2}v^2 + \frac{\gamma}{\gamma-1}\frac{p}{\rho} = \tfrac{1}{2}v_0^2 + \frac{\gamma}{\gamma-1}\frac{p_0}{\rho_0}.$$

Consider first the pressure which occurs at a stagnation point, where the fluid is brought to rest at the nose of a body. Putting $v_0 = 0$, the stagnation pressure p_0 is determined by the equation

$$\frac{p_0}{p}\frac{\rho}{\rho_0} = 1 + \frac{1}{2}\frac{\gamma-1}{\gamma}\frac{\rho v^2}{p}$$

$$= 1 + \frac{\gamma-1}{2}\frac{v^2}{c^2},$$

where c is the velocity of sound corresponding to the pressure p and density ρ of the undisturbed stream, and

$$c^2 = \frac{dp}{d\rho} = \gamma \frac{p}{\rho}.$$

Also

$$\frac{p_0}{p} \frac{\rho}{\rho_0} = \left(\frac{p_0}{p}\right)^{\frac{\gamma-1}{\gamma}},$$

and hence finally

$$p_0 = p\left(1 + \frac{\gamma-1}{2}\frac{v^2}{c^2}\right)^{\frac{\gamma}{\gamma-1}}.$$

This equation takes the place of the simpler form

$$p_0 = p + \tfrac{1}{2}\rho v^2$$

which was obtained for an incompressible fluid.

When the velocity v is small compared with the velocity of sound c, the expression for the stagnation pressure p_0 can be expanded in the series

$$p_0 = p\left(1 + \frac{\gamma}{2}\frac{v^2}{c^2} + \frac{\gamma}{8}\frac{v^4}{c^4} + \dots\right)$$
$$= p + \frac{1}{2}\rho v^2\left(1 + \frac{1}{4}\frac{v^2}{c^2} + \dots\right),$$

showing that the stagnation pressure is greater than in an incompressible fluid. Now the velocity of an aircraft is determined by a standard pressure head instrument in the form

$$v' = \sqrt{\frac{2(p_0 - p)}{\rho}},$$

and so the velocity will be over-estimated slightly if the compressibility of the air is neglected. The extent of this error is shown by table 2 and it appears that the error is less than 1 % for ordinary aeroplane speeds and is only 2 % for a speed of 300 m.p.h.

Table 2.

v/c	0·1	0·2	0·5	1·0
p_0/p	1·007	1·028	1·187	1·893
v'/v	1·001	1·005	1·032	1·129

The variation of the cross sectional area of a stream tube is determined by the equation of continuity

$$\rho v S = \text{constant},$$

which gives
$$\frac{1}{S}\frac{dS}{dv} + \frac{1}{\rho}\frac{d\rho}{dv} + \frac{1}{v} = 0.$$

Also by differentiating Bernoulli's equation
$$v + \frac{\gamma}{\gamma - 1}\left(\frac{1}{\rho}\frac{dp}{d\rho} - \frac{p}{\rho^2}\right)\frac{d\rho}{dv} = 0,$$

or
$$v + \frac{c^2}{\rho}\frac{d\rho}{dv} = 0,$$

if c is the local velocity of sound. Hence
$$\frac{dS}{dv} = -\frac{S}{v}\left(1 - \frac{v^2}{c^2}\right).$$

This equation shows that the stream tube contracts as the velocity increases if the velocity is less than the local velocity of sound, and expands if the velocity is greater than this value. It follows that the flow pattern past a body must change very considerably as the velocity approaches and exceeds the velocity of sound.

The cross-sectional area of the stream tube has a minimum value when the velocity is equal to the local velocity of sound. The characteristics at any point of the stream tube can be expressed conveniently in terms of their values at the minimum section, which will be denoted by the suffix m. The pressure, density and velocity of sound are connected by the equations
$$\left(\frac{p}{p_m}\right)^{\frac{\gamma - 1}{\gamma}} = \left(\frac{\rho}{\rho_m}\right)^{\gamma - 1} = \left(\frac{c}{c_m}\right)^2,$$

and the relationship between the velocity and the local velocity of sound, obtained from Bernoulli's equation, is
$$(\gamma - 1)\,v^2 + 2c^2 = (\gamma + 1)\,c_m{}^2.$$

Finally, the cross sectional area of the stream tube is
$$\left(\frac{S_m}{S}\right)^{\gamma - 1} = \left(\frac{\rho v}{\rho_m c_m}\right)^{\gamma - 1}$$
$$= \frac{c^2}{c_m{}^2}\left(\frac{v}{c_m}\right)^{\gamma - 1}$$
$$= \frac{\gamma + 1}{2}\left(\frac{v}{c_m}\right)^{\gamma - 1} - \frac{\gamma - 1}{2}\left(\frac{v}{c_m}\right)^{\gamma + 1}$$

These equations lead to the interesting conclusion that there is an upper limit to the velocity

$$\frac{v\,(\max)}{c_m} = \sqrt{\frac{\gamma + 1}{\gamma - 1}} = 2 \cdot 45,$$

corresponding to the condition when the pressure, density and local velocity of sound have all fallen to zero. At the other extreme, when the velocity is zero, the equations give the values

$$\frac{c_0}{c_m} = \sqrt{\frac{\gamma + 1}{2}} = 1 \cdot 095,$$

$$\frac{p_0}{p_m} = \left(\frac{\gamma + 1}{2}\right)^{\frac{\gamma}{\gamma - 1}} = 1 \cdot 893,$$

$$\frac{\rho_0}{\rho_m} = \left(\frac{\gamma + 1}{2}\right)^{\frac{1}{\gamma - 1}} = 1 \cdot 577.$$

In aeronautical problems the velocity is in general sufficiently low to justify the assumption that the air may be regarded as an incompressible fluid, but in the case of an airscrew rotating with high angular velocity, and possibly in certain other special cases, it is necessary to take account of the compressibility of the air. The compressibility may also modify the flow past a body moving with low velocity relative to the fluid, if the local velocity in any region rises to a high value.

CHAPTER III

THE STREAM FUNCTION

3·1. The determination of the flow past any body depends on the determination of the magnitude and direction of the velocity at all points of the fluid, and this velocity vector may be conveniently expressed by its three components (u, v, w) parallel to a set of orthogonal coordinate axes (x, y, z). The problem assumes a simpler form when the body is an infinite cylinder whose generators are normal to the direction of the undisturbed stream, and the flow has no component parallel to the generators. Choose the axis of z parallel to the generators of the cylinder, so that $w = 0$ at all points, and the flow will then be identical in all planes parallel to the plane $z = 0$. It is sufficient therefore to consider the flow in any plane normal to the generators of the cylinder and the problem is simplified to a motion in two dimensions only. In order to retain physical reality, this plane is assumed to have unit thickness parallel to the axis of z and curves drawn on the plane represent cylindrical surfaces of unit length in that direction.

The steady motion of a perfect fluid in two dimensions can be determined conveniently by drawing the stream lines of the motion and by the introduction of the *stream function* ψ. Take any origin O and let ψ be the flow in unit time across the curve OAP (fig. 5) joining the origin to

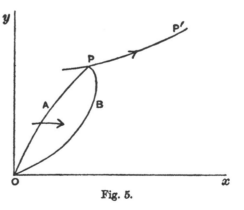

Fig. 5.

any point P of the fluid. The flow is taken to be positive in the clockwise sense about O, i.e. from left to right across the curve from the point of view of an observer at O looking along the curve towards P. The value of ψ does not depend in general on the curve which is drawn between O and P, for if OBP be another such curve, the flow across OBP must be equal to the flow across OAP unless fluid is appearing or disappearing in the region enclosed by the two curves. Hence ψ is a function of the coordinates of P, and its value will vary with the position of P. The choice of a different origin O' would merely increase the value of ψ at all points by a constant amount equal to the flow across any curve $O'O$.

Now let P' be any other point on the stream line which passes through P, and take $OAPP'$ as the curve joining P' to the origin. There is no flow across the stream line PP' and hence the flow across the curve $OAPP'$ is equal to the flow across the curve OAP, and the value of ψ at P' is the same as its value at P. It follows that the value of ψ is constant along a stream line and ψ is therefore called the *stream function*. The motion of the fluid is completely determined when the value of ψ is known as a function of the coordinates for all points of the fluid. The equation of any stream line is $\psi = C$ and the stream lines can be drawn by giving different values to the constant C. For this purpose it is best to give ψ or C values which rise by uniform increments, so that the same quantity of fluid flows between each adjacent pair of stream lines. The normal distance between adjacent stream lines is then inversely proportional to the velocity and the close approach of the stream lines in any region indicates high velocity.

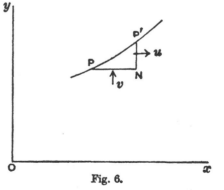

Fig. 6.

3·11. The velocity of the fluid at any point is determined simply by means of the stream function. If P and

P' are two adjacent points on different stream lines (fig. 6), the flow across the element PP' is equal to the flow across PN and NP' and can be expressed as

$$d\psi = u\,dy - v\,dx$$

in accordance with the definition of the stream function. But

$$d\psi = \frac{\partial \psi}{\partial x}\,dx + \frac{\partial \psi}{\partial y}\,dy,$$

and hence

$$u = \frac{\partial \psi}{\partial y},$$

$$v = -\frac{\partial \psi}{\partial x}.$$

In general the component of the velocity in any direction is obtained by differentiating the stream function ψ in the direction at right angles to the left. In polar coordinates, therefore, the radial and circumferential velocity components are respectively

$$u' = \frac{1}{r}\frac{\partial \psi}{\partial \theta},$$

$$v' = -\frac{\partial \psi}{\partial r}.$$

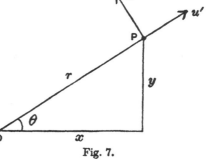

Fig. 7.

It is convenient to use both the Cartesian and the polar system of co-ordinates in many cases, and results will in general be given in both forms.

3·12. The simplest examples of the stream function correspond to uniform flow parallel to one of the coordinate axes. For a velocity U parallel to the axis of x and for a velocity V parallel to the axis of y, the stream functions are respectively

$$\psi = \quad Uy = \quad Ur\sin\theta,$$

and

$$\psi = -Vx = -Vr\cos\theta.$$

Fig. 8 shows the stream lines for the uniform flows parallel to the axes for the case $U = 1\cdot5V$, and the numbers on the

lines are the values of ψ. The broken lines are drawn through
the points which give a constant value to the sum of the two

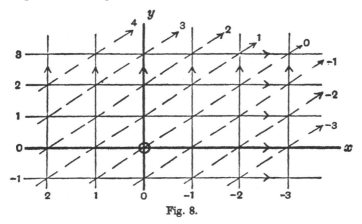

Fig. 8.

stream functions and represent a uniform flow inclined to
the coordinate axes. Two stream functions can always be
added in this manner, either analytically or graphically, and
the method is equivalent to combining the velocity vectors
at each point of the fluid. By the combination of certain
simple flow patterns in a suitable manner it is possible to
derive a number of interesting results and in particular the
flow past a circle, from which the flow past any aerofoil
section can be derived by another analytical process.

3·2. *Sources and Sinks.*

The development of different types of flow is facilitated by
the conception of sources and sinks. A *source* is a point at
which fluid is appearing at a uniform rate, and a *sink* is a
negative source or a point at which fluid is disappearing. If
there is no disturbance to the flow the fluid will pass outwards from a source equally in all directions along the radial
lines, and if m is the strength of the source or the volume of
fluid which appears in unit time, the radial velocity at
distance r from the source will be

$$u' = \frac{m}{2\pi r}.$$

The stream lines are the lines radiating from the source and the stream function ψ will have a constant value along each of these lines. Choose any radial line OA as the stream line $\psi = 0$. Then if P be a point on the radial line at angle θ to OA, the flow across the arc AP will be $\frac{m}{2\pi}\theta$ and this is the value of the stream function for the line OP. Hence

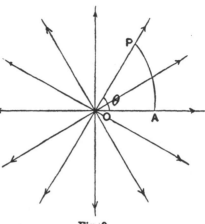

Fig. 9.

$$\psi = \frac{m}{2\pi}\,\theta = \frac{m}{2\pi}\,\text{arc}\tan\frac{y}{x}.$$

The case of a source is an exception to the general rule that the stream function at any point P has a unique value, since by choosing a curve which completely encircles the source several times it is possible to increase the value of ψ by any multiple of m. The addition of a constant to the stream function does not modify the flow pattern and the value of ψ can be made unique by the convention that θ shall always lie between the limits $\pm\,\pi$.

3·3. Source in a uniform stream.

Consider next the flow which occurs when there is a source of strength m at the origin in the presence of a uniform stream of velocity $-U$ parallel to the axis of x. The stream function for this flow is

$$\psi = -Uy + \frac{m}{2\pi}\,\theta,$$

which is the sum of the stream functions of the two separate flows. Writing $m = 2Uh$, the stream function becomes

$$\psi = U\left(h\frac{\theta}{\pi} - y\right),$$

involving the two parameters U and h. U is the velocity of the uniform stream and h is a length whose significance will appear in the course of the analysis.

The stream lines of the two separate flows are the lines parallel to the axis of x and the lines radiating from the origin respectively, and the stream lines of the combined flow can be drawn at once as the curves which pass through the points where the sum of the two stream functions has a constant value. This geometrical method is illustrated in fig. 10 for

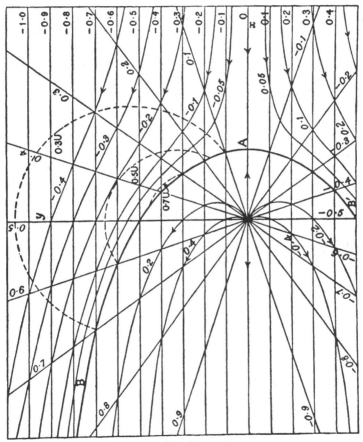

Fig. 10.

the numerical values $U = h = 1$. The stream line $\psi = 0$ consists of the positive part of the axis of x and of a curve BAB' of parabolic type. The flow from the source lies wholly within the curve BAB' and the uniform stream divides at the vertex A and flows above and below the curve. Now any stream line may be replaced by a rigid boundary without modifying the flow and an interesting interpretation of the flow is obtained if the curve xAB is taken as this boundary. The uniform stream is then passing over a level plain or sea until it meets the cliff AB, over which it is deflected in the manner indicated by the stream lines of fig. 10. With this interpretation the source no longer occurs in the region of the fluid and may be regarded simply as a mathematical device for representing the effect of the cliff.

The shape of the cliff is given by the equation $\psi = 0$ and hence

$$r \sin \theta = y = h \frac{\theta}{\pi},$$

where θ varies from 0 to π. The parameter h represents the maximum height of the cliff when r tends to infinity and θ tends to π. Thus the parameters U and h determine the velocity of the wind and the height of the cliff respectively, and the cliff is always of the same shape. Other forms, however, can be obtained by using a number of sources and sinks suitably distributed in place of the single source at the origin.

The flow may also be considered by means of the velocity components parallel to the coordinate axes. For the source these velocity components are

$$u = \frac{m}{2\pi r} \cos \theta = \frac{m}{2\pi} \frac{x}{r^2},$$

$$v = \frac{m}{2\pi r} \sin \theta = \frac{m}{2\pi} \frac{y}{r^2},$$

and hence for the flow past the cliff

$$u = -U \left(1 - \frac{h}{\pi} \frac{x}{r^2}\right),$$

$$v = \quad U \frac{h}{\pi} \frac{y}{r^2}.$$

It is now possible to determine the position of the vertex A, which is the stagnation point of the flow ($u = v = 0$). The coordinates of A are therefore

$$x = \frac{h}{\pi}, \qquad y = 0.$$

The expressions for the velocity components can also be used to determine the curves of constant vertical velocity, of constant inclination of the flow or of any other similar characteristic. The curves of constant vertical velocity v are chosen as an example, since they represent the region in which soaring flight is possible. These curves are the circles

$$x^2 + y^2 = \frac{U}{v} \frac{h}{\pi} y$$

which pass through the origin and have their centres on the axis of y. A few of these circles are drawn with broken lines in fig. 10, and these curves determine the best region for soaring flight. The maximum vertical velocity occurs on the surface of the cliff and may be determined as follows. The vertical velocity at any point is

$$v = \frac{Uh}{\pi} \frac{\sin \theta}{r},$$

and hence on the surface of the cliff

$$v = U \frac{\sin^2 \theta}{\theta},$$

which has the maximum value $v = 0\cdot725U$ at the point $\theta = 66°\cdot8$, $y = 0\cdot37h$.

This example has been discussed in some detail in order to illustrate the method of combining two flow patterns and of interpreting the result as the flow past a rigid boundary. The sources and sinks then become simple analytical devices for representing the effect of the rigid boundary, and this boundary must always be chosen to enclose all the sources and sinks.

3·4. *The method of images.*

The flow due to two sources of equal strength illustrates another analytical method of some importance. The stream lines due to two equal sources at the points A_1 and A_2 are derived very simply by the usual graphical method and are shown in fig. 11. In this case the stream lines can be shown to be hyperbolae passing through the points A_1 and A_2, but

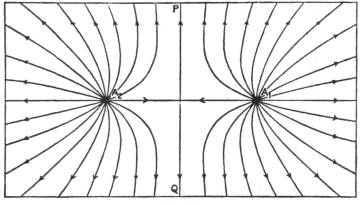

Fig. 11.

the most important feature is that the line PQ which bisects $A_1 A_2$ at right angles is a stream line and can be replaced by a rigid boundary. The stream lines to the right of the line PQ then represent the flow due to a source in the presence of a straight line boundary, and the interference of this boundary on the flow due to the source A_1 has been represented analytically by the introduction of the image A_2 of the source in the line PQ.

This method of images can be used in more complex cases. In place of the single source A_1, it is possible to take any system of sources and sinks or any closed curve representing a body. The flow past this system in the presence of a straight line boundary PQ can then be derived by introducing the image of the system in the line PQ, since the resulting flow will clearly be symmetrical about the line PQ and will have this line as one of its stream lines. The method of images can therefore be used to obtain the flow past a body near the ground.

3·5. *Source and sink in a uniform stream.*

The combination of a source and a uniform stream led to a rigid boundary extending indefinitely in one direction, but a closed curve can be obtained by using a source and sink of equal strength. Take as origin of coordinates the point midway between the source A_1 and the sink A_2, and take

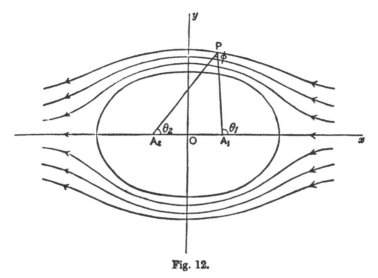

Fig. 12.

the line $A_2 A_1$ as axis of x. The stream function at any point P due to the source and sink is

$$\psi = \frac{m}{2\pi}\,(\theta_1 - \theta_2) = \frac{m}{2\pi}\,\phi,$$

where θ_1 is the angle xA_1P, θ_2 is the angle xA_2P and ϕ is the angle A_1PA_2. The stream lines ($\psi =$ constant) are therefore the system of co-axial circles passing through the points A_1 and A_2. Also if $2s$ is the distance between the source and sink

$$\tan \theta_1 = \frac{y}{x - s},$$

$$\tan \theta_2 = \frac{y}{x + s},$$

and hence

$$\tan \phi = \tan (\theta_1 - \theta_2) = \frac{2ys}{x^2 + y^2 - s^2},$$

$$\psi = \frac{m}{2\pi} \text{ arc } \tan \frac{2ys}{x^2 + y^2 - s^2}.$$

Now superimpose a uniform flow of velocity $-U$ parallel to the axis of x, and the stream function of the combined flow will be

$$\psi = -Uy + \frac{m}{2\pi} \text{ arc } \tan \frac{2ys}{x^2 + y^2 - s^2}.$$

The stream lines of this flow can be obtained by the usual graphical method and are shown in fig. 12 for a typical case. The stream line $\psi = 0$ consists of the axis of x, excluding the segment $A_1 A_2$, and of an oval curve which may be regarded as a rigid boundary. The equation of this oval curve is

$$x^2 + y^2 - s^2 = 2ys \cot \frac{2\pi U}{m} y.$$

The length b of the semi-minor axis of the oval curve is obtained at once from this equation as the value of y when x is zero, and hence

$$b^2 - s^2 = 2bs \cot \frac{2\pi U}{m} b,$$

which can be reduced to the simpler form

$$\frac{b}{s} = \cot \frac{\pi U b}{m} = \cot \frac{\pi U s}{m} \cdot \frac{b}{s}.$$

The length a of the semi-major axis is determined by the condition that the point $(a, 0)$ is a stagnation point of the flow. Now at this point the velocity is the sum of the uniform velocity $-U$ and of the components due to the source and sink, and hence

$$u = -U + \frac{m}{2\pi} \left(\frac{1}{a-s} - \frac{1}{a+s} \right)$$

$$= -U + \frac{ms}{\pi (a^2 - s^2)}.$$

At the stagnation point $u = 0$, and hence

$$\frac{a^2}{s^2} = 1 + \frac{m}{\pi U s}.$$

The shape of the oval curve depends on the single parameter Us/m and table 3 shows the relationships between the various quantities. The calculations are made by starting with a suitable series of values of Ub/m, and it appears that

Table 3.

Ub/m	Us/m	a/s	b/s	a/b
0·4	1·231	1·122	0·325	3·45
0·3	0·413	1·331	0·727	1·83
0·2	0·145	1·786	1·376	1·30
0·1	0·032	3·285	3·078	1·07

the ratio of the lengths of the axes of the oval curve tends to the limit unity as the parameter Us/m tends to zero. This limiting condition corresponds to the case when the source and sink approach indefinitely close to one another.

3·6. *Circular cylinder.*

Consider the case when the source and sink approach one another while the product of the source strength and the distance separating source and sink retains a constant finite value. Writing
$$\mu = 2ms,$$
the stream function for the source and sink is
$$\psi = \frac{\mu}{4\pi s} \arctan \frac{2ys}{x^2 + y^2 - s^2},$$
and as s tends to zero, the stream function tends to the limit
$$\psi = \frac{\mu}{2\pi} \frac{y}{x^2 + y^2} = \frac{\mu}{2\pi r} \sin \theta.$$

This combination of a source and a sink, for which s tends to zero while μ remains finite, is called a *doublet* of strength μ, and the line joining the sink to the source is called the axis of the doublet. The stream lines due to a doublet are the circles which pass through the doublet and are tangential to its axis.

Now superimpose on this flow a uniform stream of velocity

$- U$ parallel to the axis of x, and the stream function of the combined flow will be

$$\psi = - Uy + \frac{\mu}{2\pi}\frac{y}{x^2 + y^2}.$$

The stream line $\psi = 0$ consists of the axis of x and the circle

$$x^2 + y^2 = \frac{\mu}{2\pi U}.$$

Writing $\qquad\qquad \mu = 2\pi a^2 U,$

the stream function becomes

$$\psi = - Uy\left(1 - \frac{a^2}{r^2}\right) = - U\left(r - \frac{a^2}{r}\right)\sin\theta,$$

and represents the flow past a circular cylinder of radius a with centre at the origin of coordinates in a uniform stream U parallel to the axis of x in the negative sense. The stream lines of this flow can be obtained by the usual graphical method and are shown in fig. 13.

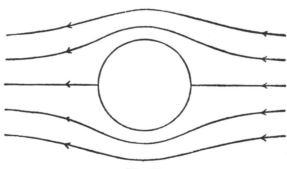

Fig. 13.

The velocity at any point is expressed most conveniently in polar coordinates, and the radial and circumferential components are respectively

$$u' = \frac{1}{r}\frac{\partial\psi}{\partial\theta} = - U\left(1 - \frac{a^2}{r^2}\right)\cos\theta,$$

$$v' = -\frac{\partial\psi}{\partial r} = U\left(1 + \frac{a^2}{r^2}\right)\sin\theta.$$

On the circumference of the circular cylinder the radial component u' is zero and the circumferential component v' is
$$v' = 2U \sin \theta,$$
which has the maximum value $2U$ when $\theta = \dfrac{\pi}{2}$.

The pressure at any point of the fluid is given by Bernoulli's equation as
$$p = p_0 + \tfrac{1}{2}\rho U^2 - \tfrac{1}{2}\rho (u'^2 + v'^2),$$
and on the circumference of the circular cylinder
$$p = p_0 + \tfrac{1}{2}\rho U^2 (1 - 4 \sin^2 \theta).$$
The pressure is symmetrical with respect to the axes of x and y, and hence there can be no resultant force on the cylinder due to the pressure distribution over its surface. This conclusion is in conflict with actual experience and fig. 14 shows

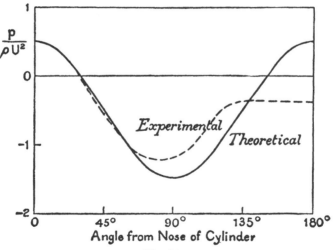

Fig. 14.

the theoretical pressure distribution compared with that given by an experimental determination on a fairly large scale*. The observed and calculated pressure distributions

* G. I. Taylor, "Pressure distribution round a cylinder," *RM*, 191, 1916. The results shown in the figure refer to a cylinder of 0·5 ft. diameter at a speed of 55 f.p.s.

agree over the front of the cylinder but are widely different in the rear. This discrepancy is due to the flow breaking away from the surface of the cylinder and forming a wake of eddying motion (see chapter VIII). The theoretical solution is of importance, however, as the basis from which the flow past an aerofoil is derived by a suitable analytical transformation.

CHAPTER IV

CIRCULATION AND VORTICITY

4·1. *Circulation.*

The analysis of the preceding chapter led to the determination of the theoretical flow past a circular cylinder in a uniform stream, but another type of flow is possible in which the fluid circulates round the cylinder. The simplest form of circulating flow is that in which the velocity has no radial component at any point while the circumferential component v' is independent of the angular position θ and depends solely on the radial distance r. By considering the equilibrium of a small element of fluid, it appears that

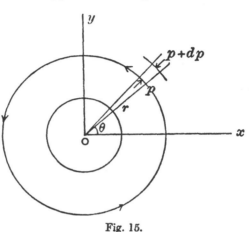

Fig. 15.

$$dp = \rho \frac{v'^2}{r}\, dr,$$

in order that the pressure on the boundary of the element shall balance the centrifugal force. If, in addition, the total pressure head H in Bernoulli's equation

$$p + \tfrac{1}{2}\rho v'^2 = H$$

is to have the same value throughout the fluid, it is necessary that the product $v'r$ shall have a constant value and this condition determines the fundamental type of circulating motion.

To determine the stream function of this circulating motion round a circular cylinder there are the equations

$$\frac{1}{r}\frac{\partial\psi}{\partial\theta} = u' = 0,$$

$$-\frac{\partial\psi}{\partial r} = v' = \frac{K}{2\pi r},$$

where K is a constant. Hence it follows that the stream function is

$$\psi = -\frac{K}{2\pi}\log r.$$

The stream lines of this flow are the circles concentric with the circular cylinder and the integral of the velocity taken round the circumference of any of these stream lines has the constant value K, which is called the circulation of the flow. More generally the *circulation* round any closed curve is defined as the integral of the tangential velocity component taken round the curve. If q is the resultant

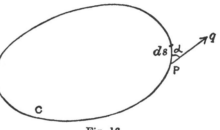

Fig. 16.

velocity at any point P of the closed curve C and if α is the angle between the direction of the velocity q and the element ds of the curve at P, then the circulation K round the curve C is

$$K = \int_C q\cos\alpha.ds.$$

The circulation will be regarded as positive in the counter-clockwise sense. For the special type of circulating motion defined by the stream function

$$\psi = -\frac{K}{2\pi}\log r,$$

the circulation has the value K for all curves enclosing the cylinder and is zero for all other curves (cf. 4·33).

4·2. *Circular cylinder with circulation.*

If the circulating flow is superimposed on the uniform flow past a circular cylinder (3·6), the stream function becomes

$$\psi = -Uy\left(1 - \frac{a^2}{r^2}\right) - \frac{K}{2\pi}\log r$$

$$= -U\left(r - \frac{a^2}{r}\right)\sin\theta - \frac{K}{2\pi}\log r,$$

and the form of the stream lines for a comparatively small value of K is as shown in fig. 17. The effect of the circulation is to increase the velocity above the cylinder and to decrease the velocity below it. In consequence there is a reduction of

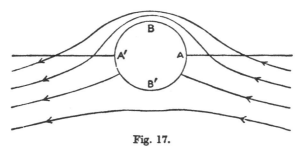

Fig. 17.

pressure above the cylinder and an increase of pressure below it, and the cylinder experiences an upward force or lift parallel to the axis of y.

The radial and circumferential components of the velocity at any point are respectively

$$u' = \frac{1}{r}\frac{\partial\psi}{\partial\theta} = -U\left(1 - \frac{a^2}{r^2}\right)\cos\theta,$$

$$v' = -\frac{\partial\psi}{\partial r} = U\left(1 + \frac{a^2}{r^2}\right)\sin\theta + \frac{K}{2\pi r},$$

and at a point on the surface of the cylinder $u' = 0$ and

$$v' = 2U\sin\theta + \frac{K}{2\pi a}.$$

The circulation causes the stagnation points to move downwards from A and A' towards B', and the two stagnation points coalesce at the point B' when the circulation K has the value $4\pi a U$. If the circulation rises above this value the flow is of the type shown in fig. 18 and there is a stagnation

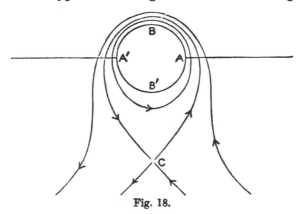

Fig. 18.

point in the fluid at C. In this case a certain part of the fluid continues to circulate round the cylinder and does not pass down stream with the general mass of the fluid.

The pressure at any point of the fluid is determined from Bernoulli's equation as

$$p = H - \tfrac{1}{2}\rho \, (u'^2 + v'^2),$$

and hence at a point on the surface of the cylinder

$$p = H - \frac{1}{2}\rho \left(2U \sin\theta + \frac{K}{2\pi a}\right)^2$$

$$= H - \frac{\rho K^2}{8\pi^2 a^2} - \frac{\rho U K}{\pi a}\sin\theta - 2\rho U^2 \sin^2\theta.$$

Now the components of the resultant force experienced by the cylinder due to the pressure distribution over its circumference are

$$X = -\int_0^{2\pi} pa \cos\theta \, d\theta,$$

$$Y = -\int_0^{2\pi} pa \sin\theta \, d\theta,$$

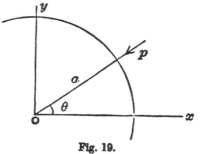

Fig. 19.

and on integration $\qquad X = 0,$

$$Y = \rho U K.$$

Thus by combining the circulation K with the uniform flow U a lift force $\rho U K$ has been obtained, and this result is of fundamental importance in the development of aerofoil theory.

4·21. Further insight into the mechanism of the lift force is obtained by examining the conditions at a great distance from the cylinder. Consider the equilibrium of the fluid contained between the surface of the cylinder and a large circle of radius r concentric with it. If q is the velocity at any point of the large circle, the pressure will be given by the equation

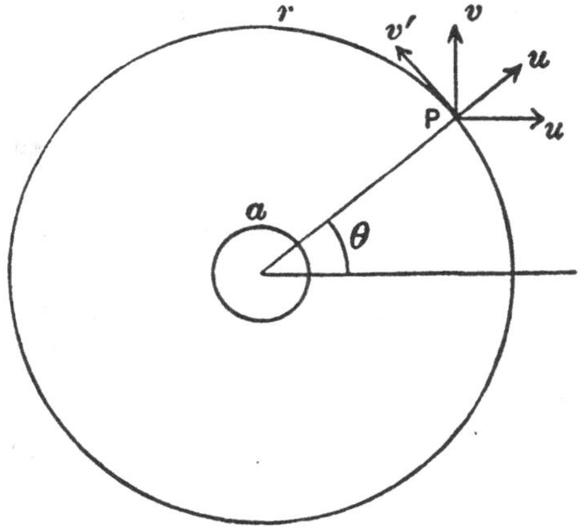

$$p = H - \tfrac{1}{2}\rho q^2,$$

Fig. 20.

and the pressure over this outer boundary will exert on the enclosed fluid a force whose components are

$$X_0 = -\int_0^{2\pi} pr \cos\theta\, d\theta = \tfrac{1}{2}\rho \int_0^{2\pi} q^2 r \cos\theta\, d\theta,$$

$$Y_0 = -\int_0^{2\pi} pr \sin\theta\, d\theta = \tfrac{1}{2}\rho \int_0^{2\pi} q^2 r \sin\theta\, d\theta.$$

Now

$$q^2 = u'^2 + v'^2 = U^2 \cos^2\theta \left(1 - \frac{a^2}{r^2}\right)^2 + \left\{U \sin\theta \left(1 + \frac{a^2}{r^2}\right) + \frac{K}{2\pi r}\right\}^2,$$

but if r is large and tends to infinity, it is sufficient to retain only the term independent of r and that proportional to $\dfrac{1}{r}$

in order to determine the force components X_0 and Y_0. To this order

$$q^2 = U^2 + \frac{UK \sin \theta}{\pi r},$$

and hence

$$X_0 = 0,$$
$$Y_0 = \tfrac{1}{2}\rho UK.$$

To these components must be added the components of the force exerted on the fluid by the pressure distribution over the surface of the cylinder, and hence the resultant force on the fluid contained between the cylinder and the large circle has the components

$$X_1 = X_0 - X = 0,$$
$$Y_1 = Y_0 - Y = -\tfrac{1}{2}\rho UK.$$

This resultant force on the fluid is normal to the direction of the undisturbed stream U and must be equal to the rate of change of momentum of the fluid. The rate at which fluid is crossing the boundary of the large circle outwards at the point P is $\rho u' r d\theta$ and the components of the momentum carried across the boundary in unit time are therefore

$$M_x = \int_0^{2\pi} \rho u' u r\, d\theta,$$

$$M_y = \int_0^{2\pi} \rho u' v r\, d\theta.$$

To the order $\dfrac{1}{r}$, the expressions for the velocity components are

$$u' = -U \cos \theta,$$
$$v' = \quad U \sin \theta + \frac{K}{2\pi r},$$
$$u = u' \cos \theta - v' \sin \theta = -U - \frac{K}{2\pi r} \sin \theta,$$
$$v = u' \sin \theta + v' \cos \theta = \frac{K}{2\pi r} \cos \theta.$$

Hence on integration

$$M_x = 0,$$
$$M_y = -\tfrac{1}{2}\rho UK,$$

which are identical with the expressions for the components of the resultant force acting on the fluid.

This analysis of the conditions at a great distance from the cylinder shows that the lift force $\rho U K$ experienced by the cylinder appears in the fluid at a great distance from the cylinder half as a change of momentum and half as the pressure distribution round the large circle*.

4·3. *Vorticity.*

The circulation round any closed curve has been defined as the integral of the tangential component of the velocity round the curve. If the curve is chosen to be a small rectangle with sides parallel to the coordinate axes, the value of the circulation is

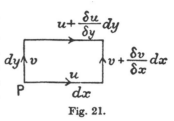

Fig. 21.

$$dK = \left(\frac{\partial v}{\partial x} - \frac{\partial u}{\partial y}\right) dx\,dy.$$

Now put

$$\omega = \frac{1}{2}\left(\frac{\partial v}{\partial x} - \frac{\partial u}{\partial y}\right),$$

and if dS is the area of the element the circulation becomes

$$dK = 2\omega\,dS.$$

In this form the equation is valid for an element of any shape and on applying it to a small circle of radius r

$$dK = 2\omega . \pi r^2 = 2\pi r . \omega r,$$

from which it follows that ω is the angular velocity of the element about its centre. Thus the value of ω at any point P of the fluid is the angular velocity of a small element surrounding the point P. The *vorticity* at any point of the fluid is defined as the value of 2ω, and the circulation round any small element is then the product of the vorticity and the area of the element. A fluid element which has vorticity is called a *vortex element* and the *strength* of a vortex element is defined as the circulation round it.

In terms of the stream function

$$u = \frac{\partial \psi}{\partial y}, \qquad v = -\frac{\partial \psi}{\partial x},$$

* See Note 2 of Appendix.

and hence the vorticity 2ω is

$$2\omega = \frac{\partial v}{\partial x} - \frac{\partial u}{\partial y} = -\left(\frac{\partial^2\psi}{\partial x^2} + \frac{\partial^2\psi}{\partial y^2}\right) = -\nabla^2\psi.$$

In polar coordinates the circulation round a small element is

$$\left\{\left(v' + \frac{\partial v'}{\partial r}\,dr\right)(r + dr)\right.$$

$$\left. - v'r\right\}\,d\theta - \frac{\partial u'}{\partial \theta}\,d\theta\,dr,$$

or $\left(\dfrac{\partial v'}{\partial r} + \dfrac{v'}{r} - \dfrac{1}{r}\dfrac{\partial u'}{\partial \theta}\right) r\,d\theta\,dr,$

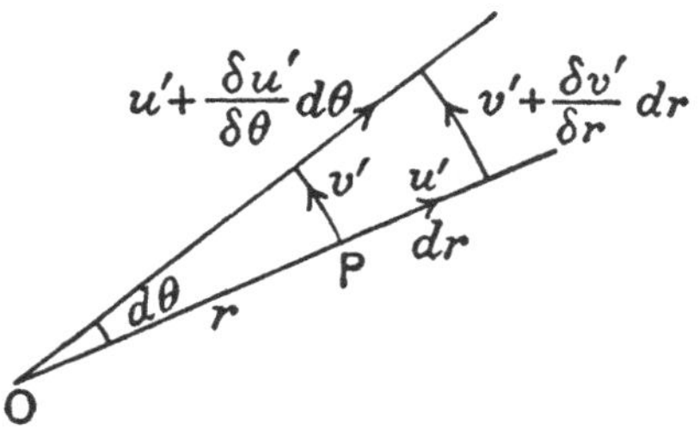

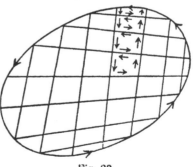

Fig. 22.

and hence

$$2\omega = \frac{\partial v'}{\partial r} + \frac{v'}{r} - \frac{1}{r}\frac{\partial u'}{\partial \theta}$$

$$= -\left\{\frac{\partial^2\psi}{\partial r^2} + \frac{1}{r}\frac{\partial\psi}{\partial r} + \frac{1}{r^2}\frac{\partial^2\psi}{\partial\theta^2}\right\},$$

or

$$\nabla^2\psi = \frac{\partial^2\psi}{\partial r^2} + \frac{1}{r}\frac{\partial\psi}{\partial r} + \frac{1}{r^2}\frac{\partial^2\psi}{\partial\theta^2}.$$

The surface enclosed by any closed curve can be divided into a large number of small elements by a double series of intersecting lines which form a network over the surface. The sum of the circulations round all these elements, taken in the positive sense, is equal to the circulation round the boundary of the surface, since the flow along any line which is common to two elements comes in twice with opposite signs and disappears from the re-

Fig. 23.

sult. There remains only the flow along the boundary of the surface which is the circulation round the closed curve. Now for any small element the circulation is equal to the product of the vorticity and the area of the element, and hence for any simple closed curve the circulation is

$$K = \int\!\!\int 2\omega\, dS = \int\!\!\int\left(\frac{\partial v}{\partial x} - \frac{\partial u}{\partial y}\right) dS,$$

where the double integral is taken over the surface enclosed by the curve. This result shows that the circulation round any closed curve is equal to the sum of the strengths of the vortices enclosed by the curve.

4·31. *Constancy of circulation and vorticity.*

The vorticity of any small element of a perfect fluid remains constant throughout the motion. The vorticity at any point P of the fluid is twice the mean angular velocity of a small element surrounding the point, and if this element is chosen as a small circle with centre at P it is evident that the pressure on the boundary of the element cannot exert any moment about the point P tending to change the angular velocity of the element. Hence, as the fluid element surrounding the point P moves with the fluid, its vorticity remains unaltered. Changes of vorticity can be produced only by tangential forces at the boundary of a fluid element and these do not occur in a perfect fluid. In a real viscous fluid tangential forces occur, particularly where the fluid is in close proximity to a rigid body, and so vorticity may arise.

Since the vorticity of the fluid elements in a perfect fluid is constant, it follows that the circulation round any closed curve moving *with* the fluid is also constant. As the curve moves with the fluid it remains continuous and unbroken and must always enclose the same fluid elements, for no fluid element can cross the curve without making a breach in its continuity. Now the circulation round the curve at any time is the sum of the strengths of the vortices enclosed within its circuit and the vorticity of all the fluid elements remains constant throughout the motion. Hence the circulation round any closed curve moving with the fluid remains constant throughout the motion.

If a closed curve moves *through* the fluid its circulation will not be constant but will be equal to the sum of the strengths of the vortices enclosed within its circuit at any moment, and the increase of circulation in any interval will be equal to the sum of the strengths of the vortices which have crossed the boundary of the curve in that interval.

4·32. *Bernoulli's equation.*

The variation of Bernoulli's constant or the total pressure head H between different stream lines is closely associated with the vorticity of the fluid. Consider a fluid element $PQQ'P'$ whose sides PQ and $P'Q'$ are elements of adjacent stream lines while PP' and QQ' are normal to them. Let $PQ = ds$, $PP' = dn$ and let R be the radius of curvature of the stream line.

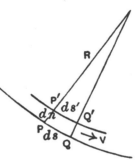

Fig. 24.

Bernoulli's equation is obtained by considering the motion along the stream line, and if V is the velocity of the fluid element

$$\rho \, ds \, dn \, V \frac{\partial V}{\partial s} = -\frac{\partial p}{\partial s} \, ds \, dn,$$

or

$$\frac{\partial p}{\partial s} + \rho V \frac{\partial V}{\partial s} = 0,$$

and hence

$$p + \tfrac{1}{2}\rho V^2 = H,$$

where H is constant along the stream line.

Resolving also normally to the stream line to obtain the balance between pressure and centrifugal force

$$\rho \, ds \, dn \, \frac{V^2}{R} = -\frac{\partial p}{\partial n} \, dn \, ds,$$

or

$$\frac{\partial p}{\partial n} + \frac{\rho V^2}{R} = 0.$$

Now the circulation round the element is

$$2\omega \, ds \, dn = V \, ds - \left(V + \frac{\partial V}{\partial n} \, dn \right) ds',$$

where

$$\frac{ds'}{ds} = \frac{R - dn}{R},$$

from which it follows that

$$2\omega = \frac{V}{R} - \frac{\partial V}{\partial n}.$$

Hence, on eliminating the radius of curvature

$$\frac{\partial p}{\partial n} + \rho V \left(2\omega + \frac{\partial V}{\partial n}\right) = 0,$$

or

$$\frac{\partial}{\partial n}(p + \tfrac{1}{2}\rho V^2) = -2\omega\rho V,$$

i.e.

$$\frac{dH}{dn} = -2\omega\rho V.$$

This equation determines the variation of the total pressure head H, and it appears that a constant value of H implies zero vorticity and conversely.

4·33. *Irrotational motion.*

When the vorticity is zero at all points of the fluid, the motion is said to be *irrotational*, since the angular velocity of any small fluid element is zero. This type of motion is of special importance, since it has been shown that vorticity cannot arise in a perfect fluid and that if the motion is irrotational at any time it will remain so always.

In irrotational motion the total pressure head H has a constant value throughout the fluid and the stream function ψ satisfies the equation $\nabla^2\psi = 0$ at all points of the fluid. The types of motion considered in chapter III are all irrotational since the stream functions satisfy this condition.

When the vorticity is zero at all points of the fluid the circulation round any closed curve or circuit, enclosing fluid only, must be zero also, but the case of a circuit enclosing a body requires some special attention. In developing the theory of the circulating motion round a cylinder the condition was imposed that the total pressure head H had a constant value throughout the fluid and it follows therefore that the motion is irrotational. The circulation round any circuit enclosing fluid only is zero, but the circulation round any circuit enclosing the cylinder has the value K. Now suppose that the circular cylinder is replaced by fluid rotating with the uniform angular velocity

$$\omega = \frac{K}{2\pi a^2},$$

where a is the radius of the cylinder. The fluid velocity will be continuous at the boundary of the cylinder and the motion outside the cylinder will be unaltered. The solid body, however, has been replaced by rotating fluid which has the vorticity 2ω at all points. Thus any circuit enclosing the cylinder will have the circulation

$$K = 2\omega \cdot \pi a^2,$$

which is the total vortex strength, while any circuit which does not enclose the cylinder will have zero circulation.

From this discussion it appears that it is possible to have irrotational flow past a body involving circulation of the flow round the body and that this flow will possess the following characteristics. The total pressure head H has a constant value and the vorticity is zero at all points of the fluid, the circulation is zero for all circuits enclosing fluid only and has a constant value for all circuits enclosing the body, and the stream function satisfies the equation $\nabla^2 \psi = 0$.

4·34. *Point vortices.*

The circulation round a small fluid element has been expressed in the form $K = 2\omega S,$

where S is the area of the element and ω is its mean angular velocity. The conception of a *point vortex* is obtained by imagining the area S to decrease to zero while the angular velocity ω increases and the circulation K remains constant. The strength of the point vortex is defined simply as the circulation K round it.

The stream function of a point vortex is derived at once from the circulating flow round a circular cylinder

$$\psi = -\frac{K}{2\pi} \log r.$$

This expression does not involve the radius of the cylinder

and therefore remains valid when the whole vorticity represented by the cylinder is concentrated at the centre. The
stream lines of a point vortex are the concentric circles with
the vortex as centre and the motion is irrotational at all
points of the fluid except the vortex itself.

The velocity at any point is normal to the line joining the
vortex to the point and has the magnitude $\dfrac{K}{2\pi r}$. Although
the vortex and the velocity are intimately related, neither
can be strictly described as caused by the other. The general
distribution of velocity associated with a vortex will be
called the *velocity field* of the vortex and the velocity at any
point will be called the *induced velocity* at the point due to
the vortex.

Point vortices may be used to build up more complex
flow patterns in the same manner as sources and sinks, and
any suitable stream line may then be replaced by a rigid

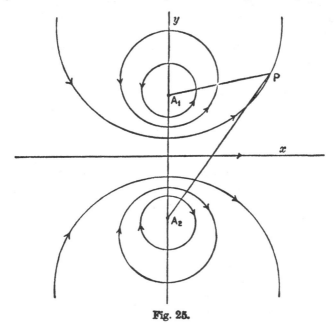

Fig. 25.

boundary. This boundary should enclose all the point vortices and the external flow will then be irrotational throughout the fluid. As an example, consider two equal vortices of opposite sign (a vortex pair) situated on the axis of y at the points $y = \pm s$. The stream function of the flow is

$$\psi = -\frac{K}{2\pi} \log \frac{A_1 P}{A_2 P},$$

and the stream lines are the co-axial circles whose limiting points are A_1 and A_2. Now proceed to a doublet, as in the case of a source and sink, by making s tend to zero while $2Ks$ retains a constant value μ. The limiting value of the stream function is

$$\psi = \frac{\mu}{2\pi} \frac{y}{x^2 + y^2},$$

which is identical with the value obtained from a source and sink (3·6). By imposing a uniform stream on a vortex pair, a series of oval bodies can be obtained, similar to those discussed in 3·5 but with their major axes normal to the stream, and on passing to the case of a doublet the flow past a circle is derived. The circulation round the circle (4·2) is obtained by adding a point vortex at the origin.

4·35. *Surface of discontinuity.*

The conception of a surface of discontinuity of velocity was introduced by Helmholtz and Kirchhoff (cf. 1·2) to explain the resultant force experienced by a body. The shape and position of the surface of discontinuity remain fixed relative to the body and the flow is tangential to the surface but the velocity has different values on the two sides of the surface. In two dimensional motion the surface of discontinuity becomes a curve of discontinuity PQ.

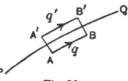

Fig. 26.

Consider a small rectangle with two sides AB and $A'B'$ of length ds parallel to an element of the curve of discontinuity and on opposite sides of it. If q and q' are the velocities on the two sides of the curve of discontinuity, the circulation round the elementary rectangle will be

$$dK = (q - q')\, ds,$$

since there is no flow along the sides AA' and BB'. The sides AA' and BB' may be made indefinitely small and it follows that the curve of discontinuity PQ must consist of a distribution of point vortices of strength $(q - q')$ per unit length. These point vortices will move with the general mass of the fluid and will have the velocity $\frac{1}{2}(q + q')$ along the stream line PQ. The velocity due to the vortices at a point adjacent to the curve of discontinuity will have equal and opposite values $\pm \frac{1}{2}(q - q')$ on opposite sides of the curve.

It follows from this discussion that a surface of discontinuity of velocity is equivalent to a vortex sheet and the distribution of point vortices which form this vortex sheet acts in the manner of roller bearings between the two fluid streams of different velocity. The type of discontinuous flow suggested by Helmholtz and Kirchhoff involves the assumption that vortex sheets spring from the sides of the body and enclose a dead-water region.

CHAPTER V

THE VELOCITY POTENTIAL AND THE POTENTIAL FUNCTION

5·1. *The Velocity Potential.*

Consider any curve OAP joining the origin O to a point P of the fluid and let ϕ be the integral of the tangential component of the velocity taken along the curve from O to P. If q is the resultant velocity at a point of the curve and if a is the angle between the direction of the velocity q and the element ds of the curve, then

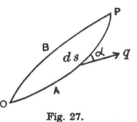

Fig. 27.

$$\phi = \int_{OAP} q \cos a\, ds.$$

In general the value of ϕ will depend on the curve connecting the points O and P, for if OBP be another such curve, the circulation round the closed curve $OAPBO$ is

$$K = \phi_{OAP} - \phi_{OBP},$$

and this circulation, in turn, is equal to the sum of the vortex strengths enclosed by the curve. In irrotational motion, however, when the vorticity is zero at all points of the fluid, ϕ has a unique value at the point P and is then called the *velocity potential*. A change of origin O will merely increase the value of ϕ by a constant amount at all points.

The integral for ϕ can be expressed in the alternative form

$$\phi = \int_O^P (u\, dx + v\, dy),$$

where u and v are the components of the velocity q measured parallel to orthogonal coordinate axes, and it follows that

$$u = \frac{\partial \phi}{\partial x}, \qquad v = \frac{\partial \phi}{\partial y}.$$

But in terms of the stream function

$$u = \frac{\partial \psi}{\partial y}, \qquad v = -\frac{\partial \psi}{\partial x},$$

and hence the velocity potential ϕ must satisfy the equation

$$\frac{\partial u}{\partial x} + \frac{\partial v}{\partial y} = 0,$$

or $\qquad \nabla^2 \phi = \dfrac{\partial^2 \phi}{\partial x^2} + \dfrac{\partial^2 \phi}{\partial y^2} = 0.$

This equation is a direct consequence of the continuity of the flow and is known as the *equation of continuity*.

The case of irrotational circulating motion round a body forms an exception to the rule that the velocity potential has a unique value at every point of the fluid. The circulation is zero for any circuit enclosing fluid only, but has a constant value K for all circuits which enclose the body once. Hence on passing round the circuit $PABP$ (fig. 28) the value of ϕ

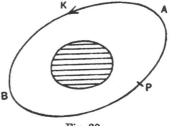

Fig. 28.

will increase by K and ϕ will be a cyclic function. This special case may be compared with the similar behaviour of the stream function ψ in the case of a source (see 3·2).

5·11. The various types of flow discussed in chapter III can be analysed in terms of the velocity potential instead of the stream function and any such flow is completely determined if either of these functions is known. The expressions for the velocity potential and stream function for the fundamental types of flow are summarised below.

Uniform flow parallel to the axis of x:

$$\phi = Ux, \qquad \psi = Uy.$$

Uniform flow parallel to the axis of y:

$$\phi = Vy, \qquad \psi = -Vx.$$

Source at the origin:

$$\phi = \frac{m}{2\pi} \log r, \qquad \psi = \frac{m}{2\pi} \theta.$$

Doublet at the origin with axis along the axis of x:

$$\phi = -\frac{\mu}{2\pi}\frac{x}{r^2}, \qquad \psi = \frac{\mu}{2\pi}\frac{y}{r^2}.$$

Point vortex at the origin:

$$\phi = \frac{K}{2\pi}\theta, \qquad \psi = -\frac{K}{2\pi}\log r.$$

Flow parallel to the negative branch of the axis of x with circulation past a circle of radius a with centre at the origin:

$$\phi = -Ux\left(1+\frac{a^2}{r^2}\right) + \frac{K}{2\pi}\theta,$$

$$\psi = -Uy\left(1-\frac{a^2}{r^2}\right) - \frac{K}{2\pi}\log\frac{r}{a}.$$

These expressions for the stream function have been developed at an earlier stage, and it can easily be verified that the corresponding expressions for the velocity potential lead to the same values of the velocity components u and v at all points and satisfy the equation of continuity.

5·12. Equipotential lines can be drawn for constant values of the velocity potential and these lines will intersect the stream lines at right angles. If dn is an element of the normal to the stream line at any point P and if $d\phi$ is the corresponding increment of the velocity potential, then the velocity of the fluid along the normal line will be $\dfrac{\partial\phi}{\partial n}$. But there is no component

Fig. 29.

of the velocity normal to a stream line by definition. Hence there can be no increment of velocity potential along the normal line and the element dn is an element of an equipotential line.

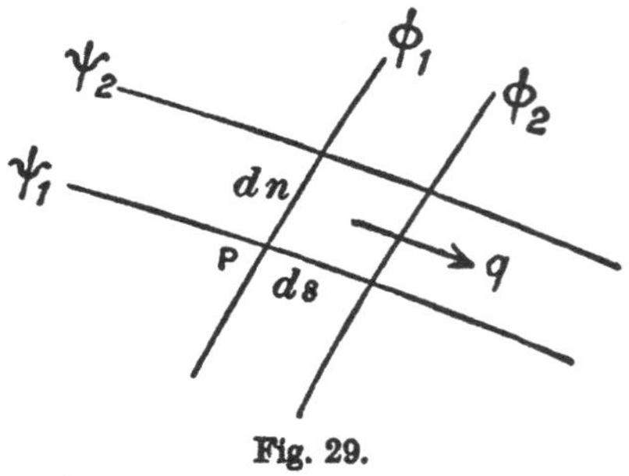

If ds is an element of the stream line and dn is an element of the equipotential line at the point P, the velocity q of the fluid is along the stream line and of magnitude

$$q = \frac{\partial \phi}{\partial s} = \frac{\partial \psi}{\partial n},$$

and if the stream lines and equipotential lines are drawn for equal increments of ψ and ϕ, the intercepts ds and dn between consecutive lines will be of equal length. It follows therefore that the stream lines and equipotential lines of any flow, drawn for equal small increments of ψ and ϕ, will divide the whole fluid region into a network of small squares. When the increments are finite, these elementary squares will be distorted and their sides will be curved but the angles of the elementary areas will remain right angles.

Fig. 30 shows the system of orthogonal lines for a source and sink at the points A_1 and A_2, the equipotential lines being

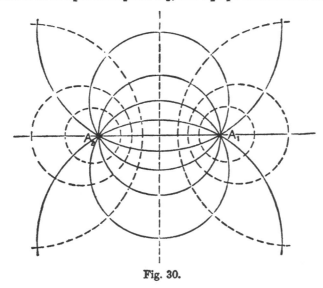

Fig. 30.

represented by broken curves. The figure may also be interpreted as the flow due to a vortex pair at the points A_1 and A_2 by interchanging the stream lines and equipotential lines. This example illustrates the general principle that any system of orthogonal lines represents two possible flows, since either set of lines may be taken to be the stream lines of the flow. It is necessary, however, to adjust the boundary conditions to fit in with the flow. Fig. 31 shows the system of orthogonal lines for a circle in a uniform stream, and, if

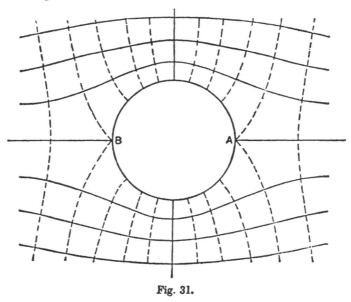

Fig. 31.

the broken lines are taken to be the stream lines, it is necessary to assume a distribution of sources and sinks over the upper and lower halves of the circumference in order to satisfy the boundary condition, since the fluid has a definite velocity normal to the circumference of the circle.

Fig. 31 also illustrates another important point. In general the stream lines and equipotential lines intersect at right angles, but this condition breaks down at the points A and B, which are the stagnation points of the flow. The proof that stream lines and equipotential lines intersect at right angles is no longer valid, the value of $\frac{\partial \phi}{\partial s}$ is zero in all directions and the stream line may turn through a sharp angle at a stagna-

tion point. It will be shown at a later stage that the equipotential line makes an equal angle with the two branches
of the stream line in this special case.

5·2. *The complex variable.*

The coordinates of a point P have been expressed either
in Cartesian or in polar co-
ordinates, but it is possible to
combine the two coordinates
required in either of these
systems into a single complex
coordinate z defined by the
equation

$$z = x + iy = r(\cos\theta + i\sin\theta),$$

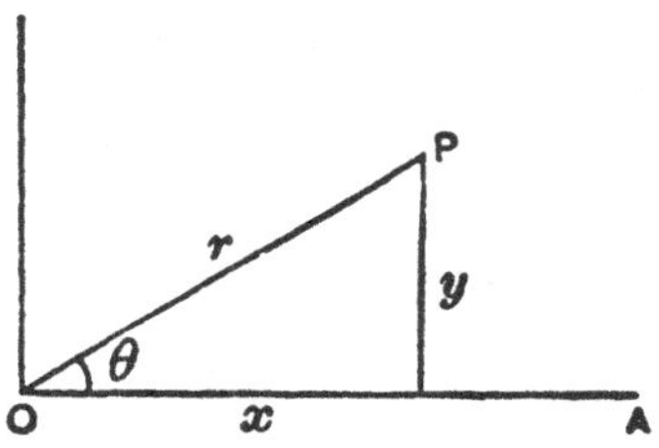

where i represents $\sqrt{-1}$ and
obeys the ordinary algebraic rules. Now

Fig. 32.

$$\frac{d}{d\theta}(\cos\theta + i\sin\theta) = -\sin\theta + i\cos\theta$$
$$= i(\cos\theta + i\sin\theta),$$

or
$$\frac{\frac{d}{d\theta}(\cos\theta + i\sin\theta)}{\cos\theta + i\sin\theta} = i,$$

and on integrating again

$$\log(\cos\theta + i\sin\theta) = i\theta,$$

or
$$\cos\theta + i\sin\theta = e^{i\theta}.$$

The complex coordinate of the point P can therefore be
expressed in the form
$$z = re^{i\theta}.$$

The coordinates (x, y) or (r, θ) define the position of the
point P relative to the origin O and the axis OA (fig. 32), but
the complex coordinate z may be interpreted more suitably
as representing the vector OP. The length of this vector is
equal to r, which is called the *modulus* of z and is written in
the alternative forms

$$r = \text{mod } z = |z|.$$

The angle θ, which defines the direction of the vector, is
called the *argument* of z.

When z is expressed in the form $(x + iy)$, x and y are called the real and imaginary parts of z respectively, and the modulus of z is equal to $\sqrt{x^2 + y^2}$. If the modulus is zero, it is evident that x and y must both be zero. Now any function $f(z)$ of the complex variable z can be separated into its real and imaginary parts, and can be expressed in the form $(X + iY)$, where X and Y are real. It follows that any complex equation $f(z) = 0$ is equivalent to the two equations obtained by equating to zero separately the real and imaginary parts of $f(z)$.

The multiplication of two complex numbers gives

$$z_1 z_2 = r_1 r_2 e^{i(\theta_1 + \theta_2)},$$

which represents a vector whose modulus is the product of the moduli and whose argument is the sum of the arguments of z_1 and z_2. Hence if any complex number or vector is multiplied by z, the length of the vector is increased by the factor $|z|$ or r and the direction of the vector is rotated through the angle θ. The factor $e^{i\theta}$ rotates a vector through the angle θ and putting $\theta = \dfrac{\pi}{2}$ it follows that the factor i rotates a vector through a right angle.

5·3. *The potential function.*

Consider any function of the complex variable z which has a single-valued differential coefficient at every point. Let

$$f(z) = \xi + i\eta,$$

and

$$\frac{df}{dz} = p + iq.$$

The differential coefficient may be expressed in the alternative forms

$$p + iq = \frac{df}{dz} = \frac{\partial f}{\partial x} = \frac{\partial \xi}{\partial x} + i\frac{\partial \eta}{\partial x}$$

$$= \frac{1}{i}\frac{\partial f}{\partial y} = -i\frac{\partial \xi}{\partial y} + \frac{\partial \eta}{\partial y},$$

and hence

$$\frac{\partial \xi}{\partial x} = \frac{\partial \eta}{\partial y} = p,$$

$$\frac{\partial \xi}{\partial y} = -\frac{\partial \eta}{\partial x} = -q.$$

Also from these last equations it follows that

$$\nabla^2 \xi = \nabla^2 \eta = 0.$$

Comparing these results with the equations which connect the velocity potential ϕ, the stream function ψ, and the velocity components u and v in any irrotational motion (5·1), it appears that ξ and η may be replaced by ϕ and ψ, and that p and q may be replaced by u and $-v$ respectively. Hence if ϕ and ψ are the real and imaginary parts of any complex function $f(z)$ they will represent possible forms of the velocity potential and stream function of an irrotational motion. It is customary to write

$$w = \phi + i\psi = f(z),$$

and then

$$\frac{dw}{dz} = u - iv.$$

The complex function w is called the *potential function* of the flow and any irrotational motion is represented completely by this function.

5·31. The fundamental types of flow summarised in 5·11 can be expressed at once in terms of the potential function, which assumes the following simple forms:

Uniform flow parallel to the axis of x:

$$w = Uz.$$

Uniform flow parallel to the axis of y:

$$w = -iVz.$$

Source at the origin:

$$w = \frac{m}{2\pi} \log z.$$

Doublet at the origin with axis along the axis of x:

$$w = -\frac{\mu}{2\pi z}.$$

Point vortex at the origin:

$$w = -i\frac{K}{2\pi} \log z.$$

Flow parallel to the negative branch of the axis of x with circulation past a circle of radius a with centre at the origin:

$$w = - U\left(z + \frac{a^2}{z}\right) - i\frac{K}{2\pi}\log\frac{z}{a}.$$

These types of flow are expressed in terms of the three simple functions z, $\dfrac{1}{z}$ and $\log z$ of the complex variable, and other types of flow can be obtained by suitable expressions for the potential function w. Consider, as an example, the flow represented by the potential function

$$w = - Uz^2 = - U\left\{(x^2 - y^2) + 2ixy\right\}.$$

The stream lines are the series of rectangular hyperbolae whose asymptotes are the axes of x and y, and by regarding these asymptotes as rigid boundaries the flow in the angle between two perpendicular walls is obtained.

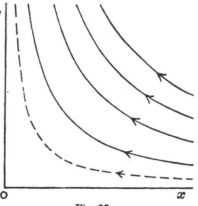

More generally, assume that the potential function is

$$w = -Uz^n = -Ur^n\,(\cos n\theta + i\sin n\theta),$$

and the irrotational flow is obtained between two

Fig. 33.

straight walls which meet at the angle $\alpha = \dfrac{\pi}{n}$. Fig. 34 illustrates the flow for $n = 4$ and $n = \frac{2}{3}$, which represent the flow in a sharp angle and round the outside of a right angle respectively.

Any complex function can be interpreted as the potential function of an irrotational motion, but the cases of practical importance are those in which the flow at a great distance from the origin approximates to a uniform stream. The potential function will then be such that, for large values of $|z|$, it can be expressed as the series

$$w = Az + B \log z + \sum_1^\infty \frac{A_n}{z^n},$$

where the coefficients A, B, A_n may be complex numbers.

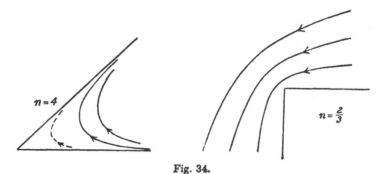

Fig. 34.

CHAPTER VI

THE TRANSFORMATION OF A CIRCLE INTO AN AEROFOIL

6·1. *Conformal transformation.*

Consider a function $f(z)$ of the complex variable z which has a unique value and a unique finite differential coefficient at every point of the z plane, and let ξ and η be the real and imaginary parts of this function:

$$\zeta = \xi + i\eta = f(z).$$

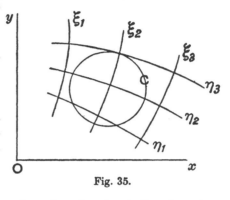

The curves of constant values of ξ and η respectively can be drawn on the z plane and will form a double series of lines intersecting at right angles, since it has been shown previously (see 5·3) that ξ and η

Fig. 35.

may represent the velocity potential and stream function of an irrotational motion.

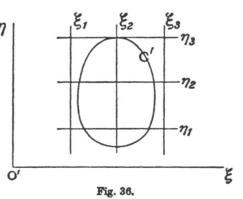

Alternatively ξ and η may be regarded as the abscissa and ordinate of a new system of coordinates, for which ζ is the complex variable, and any curve C of the z plane may be transferred to this new ζ plane. In this process the network of curved lines of the

Fig. 36.

z plane is transformed into a network of orthogonal straight lines and the curve C of the z plane will therefore appear in a distorted form C' on the ζ plane.

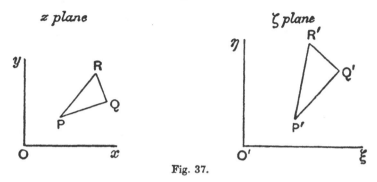

Fig. 37.

Let PQR be an elementary triangle of the z plane and let $P'Q'R'$ be the corresponding triangle of the ζ plane obtained by a transformation of this type. Also let

$$\frac{d\zeta}{dz} = f'(z) = ae^{i\alpha}.$$

Then the elementary vectors PQ (dz) and $P'Q'$ ($d\zeta$) will be related by the equation

$$d\zeta = ae^{i\alpha} dz,$$

and hence

$$|d\zeta| = a |dz|,$$

$$\arg d\zeta = \alpha + \arg dz.$$

The effect of the transformation is therefore to increase the length of the vector PQ by the factor a or $\left|\dfrac{d\zeta}{dz}\right|$ and to rotate the vector through the angle α or $\arg \dfrac{d\zeta}{dz}$. The important point, however, is that the transformation experienced by the elementary vector PQ does not depend on its direction but only on the position of the point P. It follows that the elementary triangle PQR will be transformed into a similar triangle, increased in size by the factor a and rotated through the angle α. A transformation of this type, which does not alter the shape of elementary figures, is called a *conformal transformation*.

6·11. The condition has been imposed on the function $f(z)$ that it shall have a unique value at every point of the z plane, and in consequence every point will be represented uniquely on the ζ plane. It is possible, however, that two or more points of the z plane may be represented by the same point of the ζ plane. Consider, for example, the transformation

$$\zeta = z^2,$$

which will give the same point of the ζ plane corresponding to the two points $\pm z$ of the z plane. In this case it is convenient to consider only the top half of the z plane, which will be transformed into the whole of the ζ plane. The transformation is illustrated in fig. 38, where the same letters denote corresponding points, and it will be seen that the real axis AOA' of the z plane has bent back on itself to form

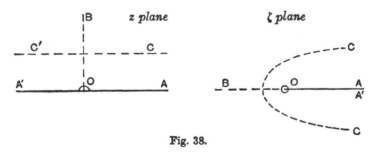

Fig. 38.

only the positive branch of the real axis of the ζ plane. It can easily be shown also that the straight lines parallel to the coordinate axes of the z plane are transformed into parabolae whose axes coincide with the real axis of the ζ plane.

The transformation function $f(z)$ may now be generalised by removing the restriction that it must have a unique value at all points of the z plane, for it is possible to consider the transformation of a limited region of the z plane into a limited region of the ζ plane, and the condition to be satisfied by the function $f(z)$ is that it shall give a unique relationship between z and ζ in these limited regions.

6·12. *Singular points.*

The simple example of conformal transformation represented in fig. 38 illustrates another important point. In a conformal transformation the angle between two intersecting lines remains unaltered after transformation, but in this particular transformation the angle $\frac{\pi}{2}$ between the lines OA and OB of the z plane has increased to π on the ζ plane. Thus the transformation has ceased to be conformal at the point O.

The ratio of elementary lengths on the two planes is $\left|\dfrac{d\zeta}{dz}\right|$, which has in general a finite value. If $\left|\dfrac{d\zeta}{dz}\right|$ is zero, a small but finite length of the z plane contracts to zero on the ζ plane, and conversely when $\left|\dfrac{d\zeta}{dz}\right|$ tends to infinity. A point at which $\left|\dfrac{d\zeta}{dz}\right|$ is zero or infinite is called a *singular point* of the transformation and at such a point the transformation ceases to be conformal.

Consider the case when $\left|\dfrac{d\zeta}{dz}\right|$ is zero at the point z_0. If ζ_0 is the corresponding value of ζ, the transformation may be written in the form

$$\zeta - \zeta_0 = (z - z_0)^n F(z),$$

where $F(z)$ does not vanish or become infinite at the point z_0 and where n is greater than unity in order that $\left|\dfrac{d\zeta}{dz}\right|$ may

z plane   *ζ plane*

Fig. 39.

be zero at this point. Now let the characteristic point of the
z plane move on the small circle

$$z = z_0 + re^{i\theta},$$

and then the corresponding variation of ζ will be given by
the equation
$$\zeta = \zeta_0 + r^n e^{in\theta} F(z_0).$$

Thus the characteristic point of the ζ plane also describes
a circular arc of small radius, but an angle θ of the z plane
corresponds to a larger angle $n\theta$ of the ζ plane.

The case when $\left| \dfrac{d\zeta}{dz} \right|$ becomes infinite can be treated in a
similar manner, and in this case an angle θ of the z plane
transforms to a smaller angle of the ζ plane.

A singular point which occurs on the boundary of the
region under consideration may be excluded by an arc of a
small circle as indicated in fig. 38, and the transformation
then becomes conformal at all points of the region. Moreover, the circular arc may be made indefinitely small and
hence in effect a singular point on the boundary of the region
will not necessarily destroy the validity of the transformation.
It is important, however, that no singular point shall occur
in the region to be transformed, and any singular point on
the boundary must satisfy certain conditions.

Consider the special case of the transformation of a circle
into an aerofoil section and assume that a singular point

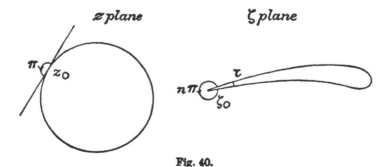

Fig. 40.

occurs on the circumference of the circle at z_0. If the transformation near this point is of the form

$$\zeta = \zeta_0 + (z - z_0)^n F(z),$$

the exterior angle π of the circle at z_0 will be transformed into the exterior angle $n\pi$ of the aerofoil at ζ_0. It is evident at once that the value of n must not exceed 2 and that to obtain a typical aerofoil shape n should be only slightly less than this value. If τ is the angle at which the upper and lower surfaces of the aerofoil meet at the trailing edge, the value of n is determined by the equation

$$\tau = (2 - n)\,\pi.$$

In the particular case $n = 2$, the aerofoil section will have a cusp at the trailing edge.

6·13. *Transformation of the flow pattern.*

The flow past any body or simple closed curve C of the z plane is determined by the potential function $w = \phi + i\psi$ and is represented by the equipotential and stream lines. These characteristic lines form an orthogonal system and after any conformal transformation of the z plane they will form an orthogonal system on the ζ plane associated with a simple closed curve C'. Hence the conformal transformation which transforms the curve C into the curve C', also transforms the flow past C into the flow past C'.

The velocity components u' and v' at any point of the ζ plane are given by the equation

$$u' - iv' = \frac{dw}{d\zeta} = \frac{dw}{dz}\frac{dz}{d\zeta} = (u - iv)\frac{dz}{d\zeta},$$

and the resultant velocities q' and q at corresponding points of the two planes are related by the equation

$$q' = q\left|\frac{dz}{d\zeta}\right|.$$

In general $\left|\dfrac{dz}{d\zeta}\right|$ has a finite value differing from zero, but at a singular point a finite velocity in one plane may correspond to an infinite velocity in the other plane. Thus in fig. 40 a finite value of the velocity q at the point z_0 will lead to an infinite velocity q' at the point ζ_0 of the aerofoil.

The irrotational flow past a circle is known and it is possible to transform a circle into any given shape of aerofoil. Hence the flow past any aerofoil section can be determined by the method of conformal transformation, and the problem of determining this flow directly may be replaced by the problem of determining the conformal transformation from the aerofoil section to a circle.

6·2. *Straight line and circle.*

An interesting and important example of the conformal transformation of a flow pattern is the application of the transformation

$$\zeta = z + \frac{a^2}{z}$$

to the circle $|z| = a$. The general point $z = re^{i\theta}$ transforms to the point whose coordinates are

$$\xi = \left(r + \frac{a^2}{r}\right)\cos\theta,$$

$$\eta = \left(r - \frac{a^2}{r}\right)\sin\theta,$$

and it follows at once that the circle $r = a$ of the z plane is transformed into the part of the real axis extending between the points $\xi = \pm 2a$.

The transformation has a simple geometrical interpretation. The complex variable z represents the vector OP of length r at angle θ to the real axis. Similarly $\dfrac{a^2}{z}$ or $\dfrac{a^2}{r}(\cos\theta - i\sin\theta)$ represents the vector OP_1 of length $\dfrac{a^2}{r}$ at angle $-\theta$ to the

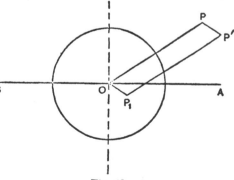

Fig. 41.

real axis, and the position of P_1 may be obtained from that of P by the double process of inversion with respect to the circle $r = a$ and reflection in the real axis. Finally, the vector OP' representing the complex variable ζ is obtained by the addition of the vectors OP and OP_1, or by completing the parallelogram POP_1P'.

By this geometrical method or by direct use of the transformation equations, the stream lines of the flow past the circle can be transformed to those of the corresponding flow past the straight line.

6·21. The potential function for uniform flow in the ζ plane parallel to the negative branch of the real axis is

$$w = -U\zeta,$$

and this represents the flow along the line AB. On transforming to the z plane, the line AB opens out to a circle and the uniform stream past this circle has the potential function

$$w = -U\zeta = -U\left(z + \frac{a^2}{z}\right).$$

Thus the method of conformal transformation gives at once a result which was obtained previously by a more tedious process.

6·22. The vertical flow past the circle can be obtained from the horizontal flow by the transformation

$$z' = iz,$$

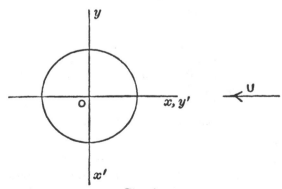

Fig. 42.

which is equivalent to rotating the axes backwards through a right angle. The potential function becomes

$$w = iU\left(z' - \frac{a^2}{z'}\right),$$

and hence the potential function for a uniform stream V parallel to the negative branch of the imaginary axis of the original system will be

$$w = iV\left(z - \frac{a^2}{z}\right).$$

6·23. By transforming the circle back to a straight line the potential function for a uniform stream normal to the line is obtained in the form

$$w = iV\sqrt{\zeta^2 - 4a^2}.$$

This result can be expressed more conveniently by means of the substitution
$$\zeta = s\sin(\lambda + i\mu),$$
where s is equal to $2a$ and is the semi-span of the line. With this substitution
$$\xi = s\sin\lambda\cosh\mu,$$
$$\eta = s\cos\lambda\sinh\mu,$$
and the periphery of the line is represented by $\mu = 0$ and $\lambda = 0$ to 2π. The potential function becomes
$$w = -Vs\cos(\lambda + i\mu),$$
and the stream function is
$$\psi = Vs\sin\lambda\sinh\mu.$$

The stream lines of this flow are shown in fig. 43 and represent the flow relative to the straight line. The flow relative to the general mass of the fluid can be derived simply by adding the vertical velocity V at every point, and the resulting flow pattern is shown in fig. 44. These stream lines represent the motion which is caused in the fluid when the line moves normal to itself with the velocity V.

6·3. *Aerofoil and circle.*

In order to obtain the flow pattern past an aerofoil it is necessary to determine the conformal transformation which converts the aerofoil section into a circle in such a manner

that the region at infinity is unaltered*. For a given aerofoil
section in the ζ plane there is a unique conformal transforma-

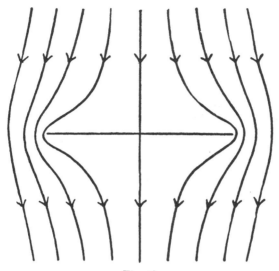

Fig. 43.

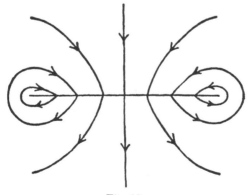

Fig. 44.

* The general theory has been developed by R. v. Mises, "Zur Theorie
des Tragflächenauftriebes," *ZFM*, 1917 and 1920.

tion which transforms the region external to the aerofoil into
the region external to a circle in the z plane, and this circle
is uniquely determined in magnitude and position. The conformal transformation is of the type

$$z = \zeta + \frac{A_1}{\zeta} + \frac{A_2}{\zeta^2} + \dots,$$

where the coefficients A_1, A_2, ... are complex numbers in
general.

Conversely a circle of the z plane can be transformed into
an aerofoil section in the ζ plane by a conformal transformation of the type

$$\zeta = z + \frac{a_1}{z} + \frac{a_2}{z^2} + \dots,$$

and by suitable choice of the coefficients $a_1, a_2, \dots$ and of the
circle it is possible to obtain any given aerofoil shape. No
limitations exist on the choice of the coefficients, but the circle
must enclose within its circumference all the singular points
of the transformation at which $\frac{d\zeta}{dz}$ is zero or infinite. The
general transformation gives

$$\frac{d\zeta}{dz} = 1 - \frac{a_1}{z^2} - \frac{2a_2}{z^3} - \dots,$$

which can become infinite only at the origin, but may be
zero at a number of points v_1, v_2, etc.

6·31. *Joukowski's hypothesis.*

The general flow past a circle contains one arbitrary
parameter, the circulation K of the flow round the circle,
and this arbitrary parameter will remain when the flow is
transformed to the flow past an aerofoil. Now an aerofoil
usually has a very small radius of curvature at the trailing
edge and in developing the theory of an aerofoil it is convenient to make the assumption that the upper and lower
surfaces of the aerofoil meet at a sharp angle at the trailing
edge. The point B of the circle which transforms into the
trailing edge of the aerofoil will then be a zero of $\frac{d\zeta}{dz}$ and if the

velocity q at the point B of the circle has a finite value, the corresponding velocity q' at the trailing edge of the aerofoil will become infinite, since

$$q' = q \left| \frac{dz}{d\zeta} \right|.$$

In order to avoid this infinite velocity at the trailing edge Joukowski suggests that the circulation K should be chosen so that the point B is a stagnation point of the flow past the circle and the velocity q is zero. The flow past the aerofoil is then such that it leaves the trailing edge tangentially and the velocity remains finite at all points.

Joukowski's hypothesis determines the circulation K uniquely when the aerofoil has a sharp trailing edge, and the aerofoil section will be assumed always to possess this characteristic. The critical discussion of Joukowski's hypothesis is reserved to a later chapter (see 9·3).

6·32. If the transformation formula is the finite series

$$\zeta = z + \frac{a_1}{z} + \frac{a_2}{z^2} + \dots + \frac{a_{n-1}}{z^{n-1}},$$

the singular points are determined from the equation

$$\frac{d\zeta}{dz} = 1 - \frac{a_1}{z^2} - \frac{2a_2}{z^3} - \dots - \frac{(n-1)a_{n-1}}{z^n}$$

$$= \left(1 - \frac{v_1}{z} \right) \left(1 - \frac{v_2}{z} \right) \dots \left(1 - \frac{v_n}{z} \right),$$

where $\Sigma v_1 = 0,$

                     $\Sigma v_1 v_2 = - a_1,$ etc.

The equation $\Sigma v = 0$ shows that the origin O of the system of coordinates has been chosen at the centroid of the singular points, but the direction of the axes is still undetermined.

Now the circle which is to be transformed into an aerofoil must enclose all the singular points within its circumference in order that the transformation shall be conformal. On the other hand the trailing edge of an aerofoil approximates to a sharp edge and to obtain this feature of the aerofoil it is necessary that one of the singular points B shall lie on the circumference of the circle. The real axis may be chosen conveniently as the line BO.

The process for obtaining an aerofoil section may now be laid down in general terms. Choose n points in the z plane which are to be the zeros of the transformation and take the origin O at the centroid of these points. Draw any circle which passes through one of the zeros, B, and en-

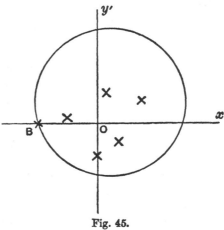

Fig. 45.

closes the remainder within its circumference. Then if BO is taken as the axis of x and if v_1, v_2, ... are the complex coordinates of the n zeros, the transformation will be

$$\frac{d\zeta}{dz} = \left(1 - \frac{v_1}{z}\right)\left(1 - \frac{v_2}{z}\right) \cdots \left(1 - \frac{v_n}{z}\right).$$

By choosing different circles and different sets of zeros of the transformation, an infinity of different aerofoil shapes can be derived. In each case the point B of the circle will transform into the trailing edge of the aerofoil and by reference to 6·12 it will be seen that the upper and lower surfaces of the aerofoil will meet in a cusp at the trailing edge.

6·33. In the most general case the transformation formula is the infinite series

$$\zeta = z + \frac{a_1}{z} + \frac{a_2}{z^2} + \cdots,$$

or

$$\frac{d\zeta}{dz} = 1 - \frac{a_1}{z^2} - \frac{2a_2}{z^3} - \cdots.$$

The circle to be transformed into an aerofoil must be such that $\frac{d\zeta}{dz}$ does not vanish or become infinite at any point outside the circle, and if the aerofoil is to have a sharp trailing edge one zero of $\frac{d\zeta}{dz}$ must lie on the circumference of the circle at the point B $(z = v)$. The transformation may then be written in the form

$$\frac{d\zeta}{dz} = \left(1 - \frac{v}{z}\right)^{n-1} f(z),$$

where $f(z)$ has a finite value other than zero at all points on and outside the circumference of the circle. In the neighbourhood of the zero B the transformation will be of the form

$$\zeta = \zeta_0 + (z - v)^n F(z),$$

and from 6·12 it follows that the upper and lower surfaces of the aerofoil will meet at the trailing edge at the angle

$$\tau = (2 - n) \pi.$$

To obtain aerofoils of conventional shape it is necessary therefore to choose n to be slightly less than 2, while if n rises to this limiting value the aerofoil has a cusp at its trailing edge.

6·4. *The Joukowski transformation.*

The simplest type of transformation formula involves two

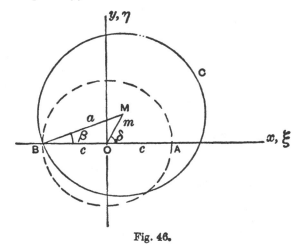

Fig. 46.

zeros A and B, and in accordance with the general theory the line joining these two points is taken as the real axis and the origin O is taken at the mid-point of AB. The coordinates of the zeros are then $z = \pm c$ and the transformation formula is

$$\frac{d\zeta}{dz} = \left(1 - \frac{c}{z}\right)\left(1 + \frac{c}{z}\right) = 1 - \frac{c^2}{z^2},$$

or

$$\zeta = z + \frac{c^2}{z}.$$

Some special applications of this transformation have been considered previously in 6·2, where it was shown that the circle on AB as diameter transforms into the part of the real axis extending between the points $\xi = \pm 2c$. More generally the transformation may be applied to any circle which encloses the points A and B within its circumference, but in order to obtain an aerofoil section with a sharp trailing edge the circle C must be chosen to pass through the point B. If the circle is slightly larger, so that the point B falls just inside the circumference, an aerofoil section is obtained with a rounded trailing edge but it is no longer possible to determine the circulation uniquely by means of Joukowski's hypothesis.

The circle C will be defined by its radius a and by the angle β between the real axis and the line joining the point B to the centre M of the circle. In order to obtain an aerofoil section of conventional shape the angle β must be small and the radius a only slightly greater than $c \sec \beta$. The position of the centre of the circle may also be specified by the length m of the line OM and by the angle δ which this line makes with the real axis. The complex coordinate of the centre M may then be expressed in the alternative forms

$$z = me^{i\delta} = ae^{i\beta} - c.$$

6·41. Circular arcs*.

Consider first the case when the centre M lies on the axis of y so that the circle C passes through both the zeros A and B and the radius of the circle is $a = c \sec \beta$.

---

* Circular arc aerofoils have been discussed by W. M. Kutta, "Auftriebskräfte in strömenden Flüssigkeiten," *Illustr. aeronaut. Mitteilungen*, 1902; "Über eine mit den Grundlagen des Flugproblems in Beziehung stehende zwei dimensionale Strömung," *Ber. d. Bayer. Akad. d. Wiss.* 1910.

The general point P of the circle has the complex coordinate $z = re^{i\theta}$ and after transformation this point becomes

$$\xi = \left(r + \frac{c^2}{r}\right)\cos\theta,$$

$$\eta = \left(r - \frac{c^2}{r}\right)\sin\theta.$$

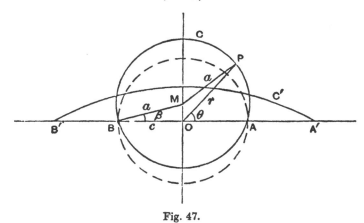

Fig. 47.

Eliminating r from these equations,

$$\xi^2 \sin^2\theta - \eta^2 \cos^2\theta = 4c^2 \sin^2\theta \cos^2\theta.$$

But from the triangle OPM

$$c^2 \sec^2\beta = a^2 = r^2 + c^2 \tan^2\beta - 2rc \tan\beta \sin\theta,$$

or $$r^2 - c^2 = 2rc \tan\beta \sin\theta,$$

and hence $$\eta = \frac{r^2 - c^2}{r}\sin\theta = 2c \tan\beta \sin^2\theta.$$

Finally, on eliminating the angle θ, the equation of the transformed curve C' becomes

$$\xi^2 + (\eta + 2c \cot 2\beta)^2 = (2c \operatorname{cosec} 2\beta)^2.$$

This is the equation of a circle, but since η has been shown to be proportional to $\sin^2\theta$, it follows that the curve C' consists only of the circular arc which lies above the real axis. The upper and lower parts of the circumference of the circle C form respectively the upper and lower surfaces of this circular arc. The end points A' and B' of the circular arc are the points $\xi = \pm\, 2c$ and the maximum ordinate is $\eta = 2c \tan \beta$, which is exactly double OM. The camber of the circular arc, defined as the maximum ordinate divided by the chord $A'B'$, is therefore $\frac{1}{2} \tan \beta$.

6·42. *Symmetrical aerofoils.*

If the centre M of the circle C is chosen on the axis of x and if the radius a is slightly greater than the fundamental

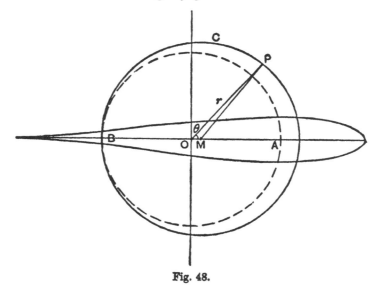

Fig. 48.

length c, the circle transforms into a symmetrical aerofoil section.

Writing $a = c\,(1 + \epsilon)$, where ϵ is a small quantity, the coordinate of the leading edge of the aerofoil is

$$\xi = c\,(1 + 2\epsilon) + \frac{c}{1 + 2\epsilon} = 2c\,(1 + 2\epsilon^2 + \ldots),$$

and as the trailing edge of the aerofoil is the point $\xi = -\,2c$, the chord of the aerofoil is $4c\,(1 + \epsilon^2)$ to a close approximation and for most purposes it is sufficiently accurate to neglect the square of ϵ and to take the chord to be $4c$.

At the general point P of the circle

$$a^2 = r^2 + (a - c)^2 - 2r (a - c) \cos \theta,$$

and retaining only the first power of ϵ,

$$r = c \{1 + \epsilon (1 + \cos \theta)\}.$$

Hence $\qquad \xi = \left(r + \dfrac{c^2}{r}\right) \cos \theta = 2c \cos \theta,$

$$\eta = \left(r - \dfrac{c^2}{r}\right) \sin \theta = 2c\epsilon (1 + \cos \theta) \sin \theta.$$

The form of the symmetrical aerofoil may be constructed by means of these equations. The thickness of the aerofoil at the centre is equal to twice the value of η when $\theta = \dfrac{\pi}{2}$ and hence $\qquad\qquad\qquad t_c = 4c . \epsilon.$

Also the maximum thickness occurs where $\cos \theta = \frac{1}{2}$, i.e. at the point which is one-quarter of the chord from the leading edge, and has the value

$$t_{\max} = 4c . \frac{3\sqrt{3}}{4} \epsilon.$$

The straight line of length $4c$ considered in 6·2 may be regarded as the centre line or skeleton of the symmetrical aerofoils. The thickness of the aerofoils is proportional to ϵ and a value $\epsilon = 0·1$ gives a maximum thickness of $0·13$ times the chord. This value is not often exceeded in practice and hence the neglect of ϵ^2 in the expression for the chord will in general give an error of less than 1%.

6·43. *Joukowski aerofoils*[*].

In general the centre M of the circle C must be taken as in fig. 46 or fig. 49. Now if BM cuts the axis of y at M_0, the circle C_0 with centre M_0 and radius M_0B will transform into a circular arc while the circle C transforms into an aerofoil. The circular arc will be the centre line or skeleton of the aerofoil, and the aerofoil may be regarded as one of the symmetrical type whose skeleton has been bent into a circular

---

[*] This type of aerofoil was introduced by Joukowski, "Über die Konturen der Tragflächen der Drachenflieger," *ZFM*, 1910.

arc of camber $\frac{1}{2}\tan\beta$. The thickness of the aerofoil will be
proportional to the
length M_0M, and
the shape of a
Joukowski aerofoil
will therefore de-
pend on the two
parameters β and
$\dfrac{a}{c}$, which determine respectively
the camber of the
centre line and the
thickness of the
aerofoil.

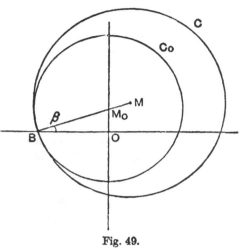

Fig. 49.

The shape of a
Joukowski aerofoil
can be obtained by
a simple geometrical construction*. The method of deriving
the point P' corresponding to any point P has been developed

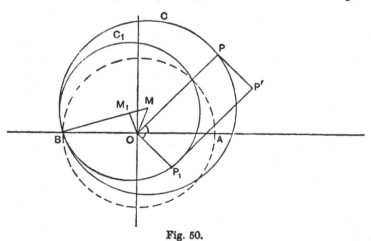

Fig. 50.

* E. Trefftz, "Graphische Konstruktion Joukowskischer Tragflächen,"
ZFM, 1913.

in 6·2. A subsidiary point P_1 is first obtained by inversion with respect to the circle on AB as diameter and by reflection in AB, and P' is then obtained by completing the parallelogram POP_1P'. Now the inverse of the circle C is another circle with centre on the line MO produced, and after reflection the centre of the auxiliary circle C_1 will lie on the line which is the reflection of OM in AB or in the axis of y. By considering the conditions near the point B it also follows that the auxiliary circle C_1 must touch the original circle C at B, and hence the centre M_1 of the auxiliary circle is the point of the line BM such that OM and OM_1 make equal angles with the axis of y.

Corresponding points P and P_1 on the circles C and C_1 are now obtained by drawing lines from the origin O at equal angles on opposite sides of the axis of x, and the point P' of the aerofoil is obtained by completing the parallelogram POP_1P'. The form of the aerofoil can be obtained by this method by taking a suitable number of points on the circumference of the circle C.

6·5. The general transformation.

The Joukowski transformation involves two zeros and leads to a doubly infinite series of aerofoils. A more general transformation formula, involving three or more zeros, leads to a greater variety of aerofoils and the types which can be derived in this manner have been discussed by R. v. Mises[*] and W. Müller[†]. This type of transformation, however, leads essentially to aerofoils which have a cusp at the trailing edge, and a more important generalisation of the Joukowski transformation is that which leads to an aerofoil section whose upper and lower surfaces meet at a finite angle at the trailing edge.

The Joukowski transformation

$$\zeta = z + \frac{c^2}{z}$$

---

[*] "Zur Theorie des Tragflächenauftriebes," *ZFM*, 1920.
[†] "Zur Konstruktion von Tragflächenprofilen," *ZAMM*, 1924.

may be written in the form

$$\frac{\zeta + 2c}{\zeta - 2c} = \left(\frac{z + c}{z - c}\right)^2,$$

and near the zero B this transformation becomes approximately

$$\zeta + 2c = -\frac{(z + c)^2}{c}.$$

In order to obtain a finite angle τ at the trailing edge of the aerofoil the transformation must be of the form

$$\zeta - \zeta_0 = (z - z_0)^n F(z)$$

in that region (cf. 6·12) and n must have the value

$$n = 2 - \frac{\tau}{\pi}.$$

This form is obtained by generalising the Joukowski transformation in the form

$$\frac{\zeta + nc}{\zeta - nc} = \left(\frac{z + c}{z - c}\right)^n.$$

This transformation has the two zeros $z = \pm c$, but the

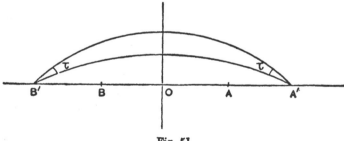

Fig. 51.

skeleton of the aerofoils is now formed by two circular arcs[*] which meet at the angle τ and the chord of the aerofoils is

---

[*] The double circular arc as the skeleton of an aerofoil was suggested by W. M. Kutta, "Über ebene Zirkulationströmungen," *Ber. d. Bayer. Akad. d. Wiss.* 1911. The transformation has been investigated by T. v. Karman and E. Trefftz, "Potentialströmung um gegebene Tragflächenquerschnitte," *ZFM*, 1918, and by W. Müller, "Zur Konstruktion von Tragflächenprofilen," *ZAMM*, 1924.

2*nc*. The transformation may also be written as an infinite series of which the first terms are

$$\zeta = z + \frac{n^2 - 1}{3}\frac{c^2}{z} + \dots.$$

There is no simple geometrical construction for aerofoils of this type and the calculation of the shape of even a symmetrical aerofoil is rather complex*. The aerofoils of this

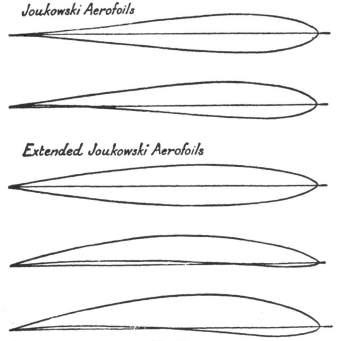

Joukowski Aerofoils

Extended Joukowski Aerofoils

Fig. 52.

generalised Joukowski type involve three arbitrary parameters, determining respectively the camber, thickness and trailing edge angle, and a wide range of aerofoil sections can be designed by this method, which are suitable for use as aeroplane wings. Some typical aerofoil sections of the Joukowski and extended types are shown in fig. 52.

* For details of the method of calculation see H. Glauert, "A generalised type of Joukowski aerofoil," *RM*, 911, 1924.

CHAPTER VII

THE AEROFOIL IN TWO DIMENSIONS

7·1. *General formulae for lift and moment.*

When the potential function w of the flow past any body is known in terms of the complex variable z, it is possible to obtain simple analytical expressions for the force and moment acting on the body. Consider the motion of the fluid contained between the surface of the body and any simple closed curve C surrounding the body. If X and Y

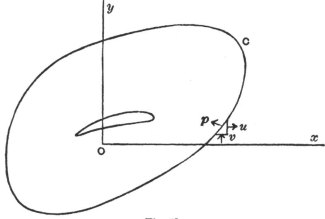

Fig. 53.

are the components of the resultant force acting on the body, the fluid will experience an equal and opposite reaction from the surface of the body in addition to the pressure which acts normally to the bounding curve C. These force components balance the rate of increase of momentum of the fluid passing out of the region under consideration, and hence

$$- X - \int_C p\,dy = \int_C \rho u\,(u\,dy - v\,dx),$$

$$- Y + \int_C p\,dx = \int_C \rho v\,(u\,dy - v\,dx),$$

where the integrals are taken round the perimeter of the curve C.

If the motion is irrotational, the total pressure head will have a constant value H at all points of the fluid and the pressure at any point will be

$$p = H - \tfrac{1}{2}\rho\, (u^2 + v^2).$$

Then also

$$X - iY = -\int_C p\, (dy + i dx) - \int_C \rho\, (u - iv)\, (u\, dy - v\, dx)$$

$$= \tfrac{1}{2}\rho \int_C \{(u^2 + v^2)(dy + i dx) - 2\, (u - iv)(u\, dy - v\, dx)\}$$

$$= \tfrac{1}{2}\rho \int_C (u^2 - v^2 - 2iuv)\, (i dx - dy).$$

But

$$\frac{dw}{dz} = u - iv,$$

and so finally

$$X - iY = \tfrac{1}{2}\rho i \int_C \left(\frac{dw}{dz}\right)^2 dz.$$

The moment about the origin of the resultant force acting on the body can be determined in a similar manner by considering the rate at which angular momentum is passing out of the region. If M_0 is the moment on the body, the equation for the motion of the fluid is

$$- M_0 + \int_C p\, (x\, dx + y\, dy) = \int_C \rho\, (vx - uy)\, (u\, dy - v\, dx),$$

and hence

$$M_0 = - \tfrac{1}{2}\rho \int_C (u^2 + v^2)\, (x\, dx + y\, dy) - \rho \int_C (vx - uy)\, (u\, dy - v\, dx)$$

$$= - \tfrac{1}{2}\rho \int_C \{(u^2 - v^2)\, (x\, dx - y\, dy) + 2uv\, (y\, dx + x\, dy)\}.$$

But

$$\int_C \left(\frac{dw}{dz}\right)^2 z\, dz = \int_C (u^2 - v^2 - 2iuv)\, (x + iy)\, (dx + i dy),$$

and the real part of this integral is identical with the integral which occurs in the expression for M_0. Hence

$$M_0 = - \tfrac{1}{2}\rho R \int_C \left(\frac{dw}{dz}\right)^2 z\, dz.$$

These integral expressions for the force and moment about the origin are valid for any number of bodies enclosed within the curve C and may be evaluated for any simple curve surrounding the bodies. Expressing the square of the differential of the potential function as the series

$$\left(\frac{dw}{dz}\right)^2 = A_0 + \frac{A_1}{z} + \frac{A_2}{z^2} + \dots$$

for large values of z, the values of the integrals become at once

$$X - iY = \tfrac{1}{2}\rho i\,(2\pi iA_1) = -\pi\rho A_1$$

and

$$M_0 = -\tfrac{1}{2}\rho R\,(2\pi iA_2) = -\pi\rho R\,(iA_2),$$

as may be verified by choosing as the curve C a circle of large radius with centre at the origin of coordinates. In these final expressions, the coefficients A_1 and A_2 will be complex quantities in general and so X, Y and M_0 will all have finite values.

7·2. Lift and moment of an aerofoil.

In order to apply this method of calculation to an aerofoil which has been derived from a circle by the conformal transformation $\zeta = f(z)$, it is necessary in the first place to determine the potential function of the flow past the circle. In the general case the origin of coordinates O is chosen at the centroid of the zeros of $\dfrac{d\zeta}{dz}$ and the circle encloses all the

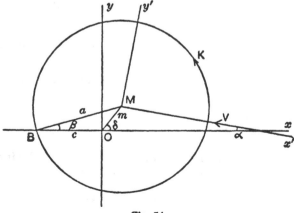

Fig. 54.

zeros except one within its circumference. The remaining zero B lies on the circumference of the circle and is transformed into the trailing edge of the aerofoil. The real axis is chosen to pass through the point B, whose complex coordinate is then taken to be $z = -c$. The circle will be assumed to be of radius a and to have its centre at the point

$$z = -c + ae^{i\beta} = me^{i\delta},$$

as indicated in the figure.

Now suppose that the undisturbed flow is of velocity V inclined at angle a to the negative direction of the real axis, and that the circulation K is chosen in accordance with Joukowski's hypothesis (see 6·31) to be such that the rear stagnation point of the flow occurs at the point B of the circle. In terms of the complex coordinate z', with origin at the centre of the circle and real axis opposed to the direction of the stream V, the potential function of the flow past the circle is (from 5·31)

$$w = -V\left(z' + \frac{a^2}{z'}\right) - \frac{iK}{2\pi}\log\frac{z'}{a},$$

and the coordinate of B is $z' = -ae^{i(a+\beta)}$.

For this flow

$$\frac{dw}{dz'} = -V\left(1 - \frac{a^2}{z'^2}\right) - \frac{iK}{2\pi z'},$$

and the circulation K must be determined so that this expression vanishes at the stagnation point B. Hence

$$V\{1 - e^{-2i(a+\beta)}\} = \frac{iK}{2\pi a}e^{-i(a+\beta)},$$

which leads to the value of the circulation

$$K = 4\pi a V \sin(a + \beta).$$

7·21. The aerofoil is derived from the circle by the general conformal transformation

$$\zeta = z + \frac{a_1}{z} + \frac{a_2}{z^2} + \cdots,$$

where the coefficients are complex in general. To determine the force on the aerofoil, it is necessary to obtain the value of $\dfrac{dw}{d\zeta}$ for the flow. Now the variables z and z' are related by the equation

$$z' = (z - me^{i\delta})\,e^{i\alpha},$$

and so

$$\frac{dw}{d\zeta} = \frac{dw}{dz'} \cdot \frac{dz'}{dz} \cdot \frac{dz}{d\zeta}$$

$$= \frac{\left(-V + V\dfrac{a^2}{z'^2} - \dfrac{iK}{2\pi z'}\right)e^{i\alpha}}{\left(1 - \dfrac{a_1}{z^2} - \dfrac{2a_2}{z^3} - \ldots\right)}.$$

Substituting for z' in terms of z and expanding in descending powers of z, this expression becomes

$$\frac{dw}{d\zeta} = -Ve^{i\alpha} - \frac{iK}{2\pi}\cdot\frac{1}{z} + \left(a^2Ve^{-i\alpha} - a_1Ve^{i\alpha} - \frac{iK}{2\pi}me^{i\delta}\right)\frac{1}{z^2} + \ldots,$$

from which it follows that

$$\left(\frac{dw}{d\zeta}\right)^2 = A_0 + \frac{A_1}{z} + \frac{A_2}{z^2} + \ldots,$$

where

$$A_0 = V^2e^{2i\alpha},$$

$$A_1 = \frac{iVK}{\pi}e^{i\alpha},$$

$$A_2 = 2a_1V^2e^{2i\alpha} - 2a^2V^2 + \frac{iVKm}{\pi}e^{i(\alpha+\delta)} - \frac{K^2}{4\pi^2}.$$

7·22. The general expression for the force on a body now gives for the aerofoil

$$X - iY = \tfrac{1}{2}\rho i \int_O \left(\frac{dw}{d\zeta}\right)^2 d\zeta$$

$$= \tfrac{1}{2}\rho i \int_O \left(\frac{dw}{d\zeta}\right)^2 \frac{d\zeta}{dz}\,dz$$

$$= \tfrac{1}{2}\rho i \int_O \left(A_0 + \frac{A_1}{z} + \frac{A_2}{z^2} + \ldots\right)\left(1 - \frac{a_1}{z^2} - \ldots\right)dz$$

$$= \tfrac{1}{2}\rho i\,(2\pi iA_1),$$

whence

$$X - iY = -i\rho VKe^{i\alpha},$$

or

$$\begin{cases} X = \rho VK \sin\alpha, \\ Y = \rho VK \cos\alpha, \end{cases}$$

which are the components of a force ρVK at right angles to the stream V. Thus the aerofoil experiences simply a lift force

$$L = \rho VK = 4\pi a\rho V^2 \sin (\alpha + \beta).$$

7·23. The moment of the lift force about the origin of coordinates is determined as

$$\begin{aligned}
M_0 &= -\tfrac{1}{2}\rho R \int_C \left(\frac{dw}{d\zeta}\right)^2 \zeta\, d\zeta \\
&= -\tfrac{1}{2}\rho R \int_C \left(\frac{dw}{d\zeta}\right)^2 \left(z + \frac{a_1}{z} + \frac{a_2}{z^2} + \ldots\right)\left(1 - \frac{a_1}{z^2} - \ldots\right) dz \\
&= -\tfrac{1}{2}\rho R \int_C \left(A_0 + \frac{A_1}{z} + \frac{A_2}{z^2} + \ldots\right)\left(1 - \frac{a_2}{z^3} - \ldots\right) z\, dz \\
&= -\tfrac{1}{2}\rho R\, (2\pi i A_2),
\end{aligned}$$

so that M_0 is the imaginary part of $\pi\rho A_2$. Now put

$$a_1 = b^2 e^{2i\gamma}$$

in the expression for A_2, and the value of M_0 becomes

$$M_0 = 2\pi b^2 \rho V^2 \sin 2\,(\alpha + \gamma) + \rho VKm \cos (\alpha + \delta).$$

This expression represents the moment about the origin of coordinates, and the moment about the centre of the circle can be derived at once as

$$\begin{aligned}
M_c &= M_0 - Lm \cos (\alpha + \delta) \\
&= 2\pi b^2 \rho V^2 \sin 2\,(\alpha + \gamma).
\end{aligned}$$

The value of this moment depends on the value of the complex coefficient a_1 in the transformation formula which defines ζ in terms of z. The lift force vanishes at the angle of incidence $-\beta$ and the moment then has the value

$$M_c = 2\pi b^2 \rho V^2 \sin 2\,(\gamma - \beta).$$

If the aerofoil is to have a constant position of the centre of pressure, i.e. if the line of action of the lift force always passes through a definite point, it is necessary that this moment should be zero. Hence the necessary and sufficient condition for a constant position of the centre of pressure of an aerofoil is that $\beta = \gamma$, or that the coefficient a_1 of the conformal transformation should be of the form $a_1 = b^2 e^{2i\beta}$.

7·24. The general expressions for the lift and moment assume simple forms for the Joukowski aerofoils which are derived by means of the conformal transformation

$$\zeta = z + \frac{c^2}{z}.$$

For aerofoils of small camber and thickness, the chord is approximately equal to $4a$ and the lift coefficient may therefore be taken as

$$C_L = 2\pi\,(\alpha + \beta).$$

This form is also valid for the majority of aerofoils which are not of the Joukowski type (see also 7·3) and is confirmed by the experimental evidence available. The theoretical slope of the curve of lift coefficient against angle of incidence is 2π per radian or 0·110 per degree, but the average slope determined experimentally is slightly less than this value, due to departure of the flow from the ideal form, and a slope of 6 per radian can be regarded as the normal value for a good aerofoil.

In the Joukowski transformation the coefficient a_1 has the value c^2 and the moment about the centre of the circle is

$$M_o = 2\pi c^2 \rho V^2 \sin 2\alpha.$$

The moment round the leading edge of the aerofoil is to a very close approximation

$$M = M_o - 2cL,$$

and to derive the corresponding coefficient it is sufficiently accurate to neglect the small difference in magnitude between a and c. Hence

$$C_M = \frac{\pi}{2}\alpha - \pi\,(\alpha + \beta)$$

$$= -\frac{\pi}{2}\beta - \frac{1}{4}C_L.$$

This formula also has been fully confirmed by experimental results and in general the moment coefficient of an aerofoil can be expressed with good accuracy as

$$C_M = C_{M_o} - \tfrac{1}{4}C_L,$$

where C_{M_o} is the moment coefficient at zero lift. The position of the centre of pressure, as a fraction of the chord measured from the leading edge of the aerofoil, is obtained by dividing the moment coefficient by the lift coefficient, and it follows that a large value of C_{M_0} implies a rapid movement of the centre of pressure. Also if C_{M_0} is zero, the aerofoil has a constant centre of pressure at a distance of one-quarter of the chord from the leading edge.

7·3. *Thin aerofoils.*

The preceding analysis determines the lift and moment on any aerofoil when the conformal transformation, by means of which the aerofoil is derived from a circle, is known. More generally, however, the shape of the aerofoil is known but the determination of the appropriate conformal transformation is of considerable difficulty. A method of solving this problem in the case of a thin aerofoil has been proposed by Munk*. The aerofoil is replaced by the curved line which is the mean of the upper and lower surfaces, and this curve is regarded as a small deviation from a straight line. A more convenient method, however, is that introduced by Birnbaum† and the following analysis is the result of applying the method of Fourier series to Birnbaum's conception of the aerofoil problem.

Choose the origin of coordinates at the leading edge of the

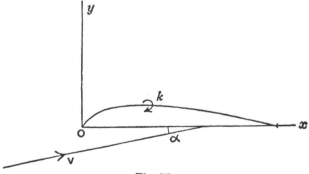

Fig. 55.

* "General theory of thin wing sections," *NACA*, 142, 1922. See also H. Glauert, "A theory of thin aerofoils," *RM*, 910, 1924.

† "Die tragende Wirbelfläche als Hilfsmittel zur Behandlung des ebenen Problems der Tragflügeltheorie," *ZAMM*, 1923.

aerofoil, with the axis of x backwards along the chord and the axis of y upwards, and consider the flow when the aerofoil is in a stream of velocity V inclined at a small angle α to the chord. There will be a circulation K round the aerofoil, corresponding to a distribution of vorticity along the surface of the aerofoil. Let $k\,dx$ be the vorticity at the element dx of the aerofoil, so that

$$K = \int_0^c k\,dx.$$

In estimating the velocity field of this system of vorticity, the approximation will be made that the vorticity is situated on the chord of the aerofoil, and then the induced velocity at the point x' of the aerofoil is determined as

$$v(x') = \int_0^c \frac{k\,dx}{2\pi(x-x')}.$$

This induced velocity is calculated for a point on the chord but may be taken to be the same as the induced velocity at the corresponding point of the aerofoil itself. The direction of the resultant velocity adjacent to the aerofoil must be parallel to the surface and so at each point of the aerofoil

$$\alpha + \frac{v}{V} = \frac{dy}{dx}.$$

These equations are sufficient to provide a complete solution of the problem in terms of the shape of the curved line which represents the aerofoil. The analysis in the general case depends on the introduction of a new coordinate θ for points of the aerofoil, defined by the relationship

$$x = \tfrac{1}{2}c\,(1 - \cos\theta),$$

so that θ varies from 0 to π along the chord of the aerofoil. It is then assumed that the vorticity k may be expressed as the series

$$k = 2V\left\{A_0 \cot \tfrac{1}{2}\theta + \sum_1^\infty A_n \sin n\theta\right\},$$

or $\qquad k\,dx = cV\left\{A_0\,(1 + \cos\theta) + \sum_1^\infty A_n \sin n\theta \sin\theta\right\}d\theta,$

where the first term represents the vorticity which occurs with a straight line aerofoil and the coefficients of the sine series depend on the shape of the aerofoil*.

* See note on p. 229.

The lift force and the moment about the leading edge of the aerofoil can be expressed simply in terms of the coefficients of this series. The lift force is

$$L = \int_0^c \rho V k\, dx$$

$$= \int_0^\pi c\rho V^2 \left\{ A_0 (1 + \cos\theta) + \sum_1^\infty A_n \sin n\theta \sin\theta \right\} d\theta$$

$$= \pi c\rho V^2 (A_0 + \tfrac{1}{2} A_1),$$

giving a lift coefficient

$$C_L = 2\pi (A_0 + \tfrac{1}{2} A_1).$$

Similarly the moment about the leading edge is

$$M = -\int_0^c \rho V k x\, dx$$

$$= -\int_0^\pi \tfrac{1}{2} c^2 \rho V^2 \left\{ A_0 (1 - \cos^2\theta) + \sum_1^\infty A_n \sin n\theta (\sin\theta - \tfrac{1}{2}\sin 2\theta) \right\} d\theta$$

$$= -\frac{\pi}{4} c^2 \rho V^2 (A_0 + A_1 - \tfrac{1}{2} A_2),$$

giving a moment coefficient

$$C_M = -\frac{\pi}{2} (A_0 + A_1 - \tfrac{1}{2} A_2)$$

$$= \frac{\pi}{4} (A_2 - A_1) - \tfrac{1}{4} C_L.$$

These expressions contain only the first three coefficients of the series for the vorticity, and the remaining coefficients correspond therefore to changes in the shape of the aerofoil which have no effect on the lift force or moment.

With the assumed value of the vorticity, the induced velocity at the point x' or θ' of the aerofoil is

$$v(x') = \frac{V}{\pi} \int_0^\pi \frac{A_0(1+\cos\theta) + \tfrac{1}{2}\sum_1^\infty A_n\{\cos(n-1)\theta - \cos(n+1)\theta\}}{\cos\theta' - \cos\theta}\, d\theta$$

$$= V\left\{ -A_0 + \tfrac{1}{2}\sum_1^\infty A_n \frac{\sin(n+1)\theta' - \sin(n-1)\theta'}{\sin\theta'} \right\},$$

since*
$$\int_0^\pi \frac{\cos n\theta}{\cos \theta - \cos \phi} d\theta = \pi \frac{\sin n\phi}{\sin \phi}$$

and the induced velocity at the point θ of the aerofoil is finally

$$\frac{v}{V} = -A_0 + \sum_1^\infty A_n \cos n\theta.$$

The condition that the flow is tangential to the surface of the aerofoil gives the relationship

$$\frac{dy}{dx} = \alpha - A_0 + \sum_1^\infty A_n \cos n\theta,$$

and then the coefficients A_n are determined from the shape of the aerofoil by evaluating the integrals

$$\alpha - A_0 = \frac{1}{\pi} \int_0^\pi \frac{dy}{dx} d\theta,$$

$$A_n = \frac{2}{\pi} \int_0^\pi \frac{dy}{dx} \cos n\theta d\theta.$$

The determination of the value of each coefficient is not necessary in general, since it is possible to obtain simple expressions for the lift force and moment about the leading edge directly in terms of the shape of the aerofoil by means of the following two integrals:

(1) $\epsilon_0 = \dfrac{2}{\pi} \int_0^\pi \dfrac{y}{c} \dfrac{d\theta}{1 + \cos \theta}$

$\qquad = \dfrac{2}{\pi} \left[\dfrac{y}{c} \sqrt{\dfrac{1 - \cos \theta}{1 + \cos \theta}} \right]_0^\pi - \dfrac{2}{\pi} \int_0^\pi \dfrac{1}{c} \dfrac{dy}{dx} \dfrac{dx}{d\theta} \sqrt{\dfrac{1 - \cos \theta}{1 + \cos \theta}} d\theta,$

and the first term vanishes if y tends to zero at the trailing edge of the aerofoil more rapidly than $\sqrt{c - x}$. Then

$$\epsilon_0 = -\frac{1}{\pi} \int_0^\pi \frac{dy}{dx} (1 - \cos \theta) d\theta = A_0 + \tfrac{1}{2} A_1 - \alpha.$$

(2) $\mu_0 = \displaystyle\int_0^\pi \dfrac{y}{c} \cos \theta d\theta = \left[\dfrac{y}{c} \sin \theta \right]_0^\pi - \int_0^\pi \dfrac{1}{c} \dfrac{dy}{dx} \dfrac{dx}{d\theta} \sin \theta d\theta$

$\qquad = -\displaystyle\int_0^\pi \dfrac{1}{4} \dfrac{dy}{dx} (1 - \cos 2\theta) d\theta$

$\qquad = -\dfrac{\pi}{4} (\alpha - A_0 - \tfrac{1}{2} A_2).$

* See note at end of chapter.

Thus in terms of the two integrals ϵ_0 and μ_0

$$C_L = 2\pi \, (\alpha + \epsilon_0),$$

$$C_M = 2\left(\mu_0 - \frac{\pi}{4}\epsilon_0\right) - \frac{1}{4}C_L,$$

and the determination of the lift and moment coefficients of any thin aerofoil has been reduced to the evaluation of the two simple integrals ϵ_0 and μ_0, and the coefficients are of the same form as for a Joukowski aerofoil (cf. 7·24).

7·31. When the form of an aerofoil is a simple analytical expression, the values of ϵ_0 and μ_0 can be obtained by direct integration. An example of some interest is the aerofoil whose form is defined by the equation

$$\frac{y}{c} = h\frac{x}{c}\left(1 - \frac{x}{c}\right)\left(1 - \lambda\frac{x}{c}\right),$$

which represents an aerofoil with reflex curvature towards the trailing edge when the value of λ lies between 1 and 2. On integration the values of ϵ_0 and μ_0 are found to be

$$\epsilon_0 = \frac{1}{8}h\,(4 - 3\lambda),$$

$$\mu_0 = \frac{\pi}{64}h\lambda,$$

and hence $\qquad C_{M_\bullet} = \frac{\pi}{32}h\,(7\lambda - 8).$

Thus an aerofoil with constant centre of pressure is obtained when λ has the value $\frac{8}{7}$.

The evaluation of the integrals in the general case is best performed by graphical methods. For this purpose the integrals are expressed in Cartesian coordinates with the aerofoil chord as unit length, and then

$$\epsilon_0 = \int_0^1 yf_1\,(x)\,dx \quad \text{and} \quad \mu_0 = \int_0^1 yf_2\,(x)\,dx,$$

where $\qquad f_1\,(x) = \dfrac{1}{\pi\,(1-x)\,\sqrt{x\,(1-x)}},$

$$f_2\,(x) = \frac{1-2x}{\sqrt{x\,(1-x)}}.$$

The numerical values of these functions at suitable points of the aerofoil chord are given in table 4 below.

The determination of μ_0 in this manner presents no difficulty since $yf_2(x)$ tends to zero at both ends of the aerofoil in general although $f_2(x)$ tends to infinity. In the case of ϵ_0, however, the value of $yf_1(x)$ generally tends to infinity at the trailing edge of the aerofoil, but this difficulty can be avoided by performing the graphical integration from the leading edge to the point $x = 0.95$ and by estimating the additional part analytically on the assumption that this last section of the aerofoil is a straight line. The additional contribution to the value of ϵ_0 can easily be shown to be $2.9y'$, where y' is the ordinate at the point $x = 0.95$.

This theoretical method of determining the angle of incidence and moment coefficient at zero lift leads to results which are in close agreement with experimental determinations of these quantities. If the experimental values are not given relative to the line joining the leading and trailing edges of the aerofoil a slight correction is required in making a comparison between theory and experiment.

Table 4.

x	$f_1(x)$	$f_2(x)$	x	$f_1(x)$	$f_2(x)$
0·025	2·09	6·10	0·50	1·27	0
0·05	1·54	4·13	0·60	1·62	−0·41
0·10	1·18	2·67	0·70	2·31	−0·87
0·20	1·00	1·50	0·80	3·98	−1·50
0·30	0·99	0·87	0·90	10·6	−2·67
0·40	1·08	0·41	0·95	29·2	−4·13

NOTE. *The value of the integral*
$$I_n = \int_0^\pi \frac{\cos n\theta}{\cos\theta - \cos\phi}\,d\theta.$$

The evaluation of this integral requires some special care, since the denominator of the integrand vanishes at the point

$\theta = \phi$ of the range of integration. It is necessary therefore to obtain the value of I_n by integrating from 0 to $(\phi - \epsilon)$ and from $(\phi + \epsilon)$ to π and by taking the limit as ϵ tends to zero.

Considering first the value of I_0,

$$\int_0^{\phi-\epsilon} \frac{d\theta}{\cos\theta - \cos\phi} = \left[\frac{1}{\sin\phi} \log \frac{\sin\frac{1}{2}(\phi+\theta)}{\sin\frac{1}{2}(\phi-\theta)} \right]_0^{\phi-\epsilon}$$

$$= \frac{1}{\sin\phi} \{\log\sin(\phi - \tfrac{1}{2}\epsilon) - \log\sin\tfrac{1}{2}\epsilon\},$$

$$\int_{\phi+\epsilon}^{\pi} \frac{d\theta}{\cos\theta - \cos\phi} = \left[\frac{1}{\sin\phi} \log \frac{\sin\frac{1}{2}(\theta+\phi)}{\sin\frac{1}{2}(\theta-\phi)} \right]_{\phi+\epsilon}^{\pi}$$

$$= \frac{1}{\sin\phi} \{\log\sin\tfrac{1}{2}\epsilon - \log\sin(\phi + \tfrac{1}{2}\epsilon)\},$$

and hence $I_0 = \lim_{\epsilon \to 0} \left\{ \frac{1}{\sin\phi} \log \frac{\sin(\phi - \frac{1}{2}\epsilon)}{\sin(\phi + \frac{1}{2}\epsilon)} \right\} = 0.$

Then $I_1 = \int_0^{\pi} \frac{\cos\theta}{\cos\theta - \cos\phi} \, d\theta$

$$= \int_0^{\pi} \left(1 + \frac{\cos\phi}{\cos\theta - \cos\phi}\right) d\theta$$

$$= \pi + I_0 \cos\phi$$

$$= \pi,$$

and more generally if $n > 1$

$$I_{n+1} + I_{n-1} = \int_0^{\pi} \frac{\cos(n+1)\theta + \cos(n-1)\theta}{\cos\theta - \cos\phi} \, d\theta$$

$$= \int_0^{\pi} \frac{2\cos\theta\cos n\theta}{\cos\theta - \cos\phi} \, d\theta$$

$$= \int_0^{\pi} \left(2\cos n\theta + \frac{2\cos\phi\cos n\theta}{\cos\theta - \cos\phi}\right) d\theta$$

$$= 2\cos\phi \, I_n.$$

The solution of this recurrence formula

$$I_{n+1} - 2\cos\phi I_n + I_{n-1} = 0,$$

with the initial conditions $I_0 = 0$ and $I_1 = \pi$, leads to the final result

$$I_n = \int_0^{\pi} \frac{\cos n\theta}{\cos\theta - \cos\phi} \, d\theta = \pi \frac{\sin n\phi}{\sin\phi}.$$

CHAPTER VIII

VISCOSITY AND DRAG

8·1. *The drag of a bluff body.*

The theory of the two-dimensional motion of a perfect fluid has led to the determination of the lift of an aerofoil by means of the assumption of a circulation of the flow, but the solution is incomplete in several respects. The conditions which cause the circulation to develop at the commencement of the motion have not been investigated and the magnitude of the circulation is indeterminate except in the case of an aerofoil with a sharp trailing edge. Joukowski's hypothesis that the circulation must be such that the flow leaves the trailing edge smoothly also requires critical examination. Finally, the theory has not indicated the existence of any drag force on the aerofoil.

To examine these problems fully it is necessary to depart from the simple assumption of a perfect fluid and to determine the effects of the viscosity or internal friction, but some insight into the drag of a body can be obtained without introducing this complication. In developing the theory of the lift force it was convenient to consider the class of bodies which give a large lift force associated with a relatively small drag force, so that the latter might be neglected without modifying the essential conditions of the problem. Similarly in examining the drag force it is convenient to consider in the first place bodies of bluff form, symmetrical about the direction of motion, so that the lift force is zero and the drag force is large. The motion will be assumed to proceed in two dimensions as before.

The simplest form of bluff body is a flat plate at right angles to the general stream, which is represented in two dimensions as a line AB of breadth b. The irrotational flow of a perfect fluid past this line is shown in fig. 43, but this type of flow gives zero drag and is unsatisfactory also because the fluid velocity becomes infinite at the edges of the plate.

An alternative type of flow was suggested by Kirchhoff and Helmholtz to overcome these difficulties and is represented in fig. 56. Curves of discontinuity of velocity are assumed to spring from the points A and B and to pass down stream enclosing a dead-water region. In consequence the

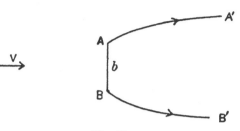

Fig. 56.

plate experiences greater pressure on the front than on the rear face, and there is a drag force*

$$D = \frac{\pi}{4 + \pi} b\rho V^2$$

corresponding to a drag coefficient

$$C_D = \frac{2\pi}{4 + \pi} = 0.88.$$

This value of the drag coefficient is approximately half that obtained from experimental determinations of the drag of a flat plate, but the conception that the flow breaks away from the surface at the edges of the plate is in accordance with fact and can be used as the basis for developing a theory of drag.

8·11. *Vortex streets†*.

The curves of discontinuity of velocity AA' and BB' which spring from the edges of the plate are essentially vortex sheets (cf. 4·35) and may be conceived as a succession of point vortices which act as roller bearings between the dead-water region and the general stream. Now a single row of equal point vortices evenly spaced along a straight line can be shown to be unstable. In the equilibrium position all the vortices will be at rest, since the induced velocity components at any vortex due to two vortices at equal distances on

* See Lamb, *Hydrodynamics*, § 76.   † See Note 3 of Appendix.

opposite sides are equal and opposite. If, however, one vortex

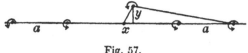

Fig. 57.

receives a small displacement (x, y) it will experience the induced velocity components

$$u = -\frac{K}{2\pi} \sum_{-\infty}^{\infty} \frac{y}{(na - x)^2 + y^2},$$

$$v = -\frac{K}{2\pi} \sum_{-\infty}^{\infty} \frac{na - x}{(na - x)^2 + y^2},$$

where K is the strength of each vortex, a is the distance between successive vortices and the summations extend over all integral values of n other than zero. For a small disturbance these expressions may be replaced by the approximations

$$u = -\frac{K}{2\pi} \sum_{-\infty}^{\infty} \frac{y}{n^2 a^2} = -\frac{\pi}{6} K \frac{y}{a^2},$$

$$v = -\frac{K}{2\pi} \sum_{-\infty}^{\infty} \frac{1}{na}\left(1 + \frac{x}{na}\right) = -\frac{\pi}{6} K \frac{x}{a^2}.$$

Hence the equations of motion of the point vortex under consideration are

$$\frac{dx}{dt} + \lambda y = 0,$$

$$\frac{dy}{dt} + \lambda x = 0,$$

where

$$\lambda = \frac{\pi}{6}\frac{K}{a^2},$$

and on eliminating y

$$\frac{d^2x}{dt^2} - \lambda^2 x = 0.$$

The solution of this differential equation is

$$x = Ae^{\lambda t} + Be^{-\lambda t},$$

which represents an unstable motion, since the first term increases indefinitely with the time and the vortex departs more and more from its equilibrium position.

The conditions behind a bluff body are more complex since

there are two vortex rows, and in order to discuss the stability
fully it is necessary to give a small disturbance to each of the
vortices. Far be-
hind the body the
vortices must lie
on two straight
lines parallel to
the general direc-
tion of motion,

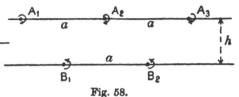

Fig. 58.

and it can easily be shown that there are only two possible
configurations. In order that the vortices may retain their
positions on the two parallel lines, the induced velocity at
any vortex must be parallel to the lines. This condition is
satisfied if any vortex B_1 of one row is exactly opposite a
vortex A_1 of the other row, or if it is opposite the mid-point
between two vortices A_1 and A_2 of the other row; each
vortex will then experience the same induced velocity u in
the sense shown in fig. 58. This velocity u is the velocity of
the vortices relative to the general mass of the fluid. For
any other configuration the induced velocity has a component
normal to the vortex rows and the configuration will not be
maintained.

The stability of these two systems has been examined by
Karman and Rubach*, and it appears that the first configuration with the vortices in pairs is essentially unstable,
but that the second configuration with alternate vortices is
stable provided the distance h between the rows and the
distance a between successive vortices of each row are related by the equation

$$\sinh \frac{\pi h}{a} = 1,$$

or $h = 0.281a.$

A double vortex row of this stable type will be called a

    * Karman, "Über den Mechanismus des Widerstandes den ein bewegter
Korper in Flüssigkeit erfahrt," *Göttingen Nachrichten*, 1911. Karman and
Rubach, "Über den Mechanismus des Flüssigkeits und Luftwiderstandes,"
Phys. Zeitschrift, 1912. The analysis is given by Lamb, *Hydrodynamics*,
§ 156.

Karman vortex street. The strength K of each point vortex will be called the strength of the street and the distance h between the rows will be called the breadth of the street. The distance a separating successive vortices of each row is a constant multiple of the breadth h, and the induced velocity u of each of the vortices is given by the equation

$$K = 2\sqrt{2}au.$$

8·12. *The form drag.*

A fully developed Karman vortex street exists far behind a bluff body but there must be an intermediate stage, shown

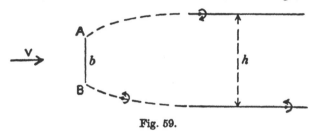

Fig. 59.

by the broken lines ot fig. 59, connecting the body and the vortex street. As the flow proceeds, the vortices pass down stream with the velocity $(V - u)$ relative to the body and new vortices must be formed alternately at the two sides of the body, which is in accordance with the observed flow past a bluff body. The frequency with which the vortices are formed at one edge of the body will be

$$f = \frac{V - u}{a}.$$

The formation of these vortices, combined with the general pressure distribution of the flow pattern, causes a drag force*

$$D = \rho K (V - 2u)\frac{h}{a} + \rho \frac{K^2}{2\pi a},$$

and on inserting the values of a and K in terms of h and u, this equation becomes

$$D = h\rho V^2 \left\{ 2\cdot83 \left(\frac{u}{V}\right) - 1\cdot12 \left(\frac{u}{V}\right)^2 \right\},$$

* Karman, *loc. cit.*

or
$$C_D = \frac{h}{b}\left\{5\cdot66\left(\frac{u}{V}\right) - 2\cdot24\left(\frac{u}{V}\right)^2\right\}.$$

The drag force due to the shedding of vortices from the sides of the body and the formation of a Karman vortex street will be called the *form drag*, to distinguish it from the drag due to tangential forces and skin friction on the surface of the body. Karman's theory appears to be in accordance with the actual conditions of flow, and if the values of u and h are determined experimentally the equation for the drag leads to a value in good agreement with the observed drag[*]. The theory is, however, incomplete and further investigation of the flow between the body and the vortex street is required to determine the values of u and h theoretically.

The form drag depends on the shape of the body. As a first rough approximation the drag coefficient may be taken to be

$$C_D = 5\cdot66\frac{h}{b}\frac{u}{V} = 0\cdot562\frac{K}{bV},$$

and to this order of approximation the drag is simply proportional to the strength of the vortices which are shed at the sides of the body. A body of bluff form, particularly if it has sharp edges like a flat plate, will shed strong vortices and will have a large form drag, but for a body of "good" shape, such as a symmetrical aerofoil section, the form drag appears to be negligibly small and the drag experienced is due mainly to the tangential forces or skin friction.

8·2. *Viscosity*.

All real fluids possess the property of internal friction or viscosity by virtue of which tangential stresses may occur at the surface of separation of two adjacent fluid elements. These tangential stresses are zero when the fluid is at rest, and in general they depend on the relative velocity of the adjacent fluid elements. The viscosity of a fluid may be defined conveniently by considering the steady motion in layers normal to the axis of y. The layer of fluid between the planes y and $(y + dy)$ will have a velocity u at all points and

[*] Karman, *loc. cit.*

u will be a function of y only. When the fluid moves
in layers in this
manner, it is said
to be in *laminar
motion*. The re-
lative velocity
of two adjacent
layers is $\frac{\partial u}{\partial y} dy$
and the tangen-
tial force at the
surface of separa-
tion is $\mu \frac{\partial u}{\partial y}$ per

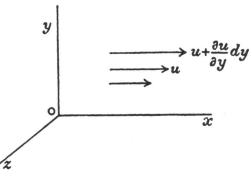

Fig. 60.

unit area, where μ is the *coefficient of viscosity* of the fluid.
This definition of the tangential force due to viscosity is based
on the conception that the frictional force depends on the
relative velocity of the adjacent fluid elements and is justified
by the accuracy of the results which can be deduced from it.

When two parallel layers of fluid are moving in the same
direction with different velocities, the surface of separation
is a vortex sheet and the elementary vortices of this sheet
act as roller bearings between the two layers of the fluid.
The tangential stress at the surface of separation is intimately
related to this vortex sheet and the work which must be
done against the tangential stress is represented by the
dissipation of energy which occurs in the vortices.

To complete the definition of the nature of a viscous fluid
it is necessary to consider the conditions at a solid boundary.
The motion of the fluid over the surface of a body will cause
a finite tangential force on the surface and it follows that the
layer of fluid immediately in contact with the surface must
be at rest relative to the surface, for if this condition were
not satisfied $\frac{\partial u}{\partial y}$ would tend to infinity at the surface and the
tangential force would also tend to infinity unless the co-
efficient of friction between solid and fluid were indefinitely
small compared with that between two fluid layers. This

condition of zero slip at a solid boundary is confirmed by experiment and by the accuracy of the results deduced from it.

The frictional force at the surface of separation of two fluid layers in laminar motion has been defined as $\mu \dfrac{\partial u}{\partial y}$ per

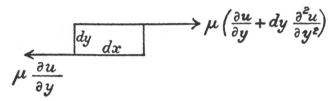

Fig. 61.

unit area, and hence the force on a fluid element of thickness dy and of area S normal to the axis of y will be

$$\mu \left(\frac{\partial u}{\partial y} + \frac{\partial^2 u}{\partial y^2}\, dy \right) S - \mu \frac{\partial u}{\partial y} S = \mu \frac{\partial^2 u}{\partial y^2} S\, dy,$$

which is $\mu \dfrac{\partial^2 u}{\partial y^2}$ per *unit volume*. It is customary, however, to work in terms of the force per *unit mass* of the fluid, and hence for the laminar flow under consideration

$$X = \frac{\mu}{\rho} \frac{\partial^2 u}{\partial y^2} = \nu \frac{\partial^2 u}{\partial y^2},$$

where ν is the coefficient of viscosity divided by the density of the fluid and is called the *kinematic coefficient of viscosity*.

8·21. *Laminar flow between flat plates.*

It is now possible to determine the laminar flow between

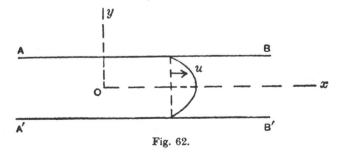

Fig. 62.

two parallel flat plates, which is the same as the two-dimensional flow in a channel between two parallel straight lines AB and $A'B'$. The equation of motion of the fluid is

$$\mu \frac{\partial^2 u}{\partial y^2} = \frac{dp}{dx},$$

which expresses the fact that the viscous force must be balanced by the pressure difference on any fluid element. The velocity u is a function of the coordinate y and the pressure p of the coordinate x.

On integrating the equation of motion

$$u = a + by + \frac{1}{2\mu} \frac{dp}{dx} y^2,$$

and if the origin O is chosen midway between the two boundaries which are at distance $2h$ apart

$$u = -\frac{1}{2\mu} \frac{dp}{dx} (h^2 - y^2).$$

The pressure decreases uniformly along the stream and the velocity distribution is parabolic across the channel. The mean velocity V of the stream is determined by the equation

$$V = \frac{1}{2h} \int_{-h}^{h} u \, dy = -\frac{h^2}{3\mu} \frac{dp}{dx},$$

and hence $$u = \tfrac{3}{2} V \left(1 - \frac{y^2}{h^2} \right).$$

The frictional drag on length l and breadth b of the two walls of the channel may be estimated from the pressure gradient as

$$D = 2hb \left(-l \frac{dp}{dx} \right) = 3\mu V \frac{S}{h},$$

where S is the "wetted" surface $2bl$. Alternatively the drag may be estimated directly from the tangential force on the surface as

$$D = \mu \left(\frac{\partial u}{\partial y} \right)_0 S,$$

where the suffix indicates that the value of $\frac{\partial u}{\partial y}$ must be taken at the boundary with dy measured into the fluid. Now

$$\frac{\partial u}{\partial y} = -\frac{3Vy}{h^2},$$

and hence

$$\left(\frac{\partial u}{\partial y}\right)_0 = \frac{3V}{h},$$

which leads to the same value of the drag as that which was obtained from the pressure gradient.

8·22. *Numerical values.*

By inspection of the formulae for the viscous force on a fluid element it can be seen that the dimensions of the two coefficients of viscosity are respectively

$$\mu \qquad ML^{-1}T^{-1},$$
$$\nu \qquad L^2T^{-1},$$

and in particular the kinematic coefficient of viscosity ν has the dimensions of a length multiplied by a velocity. In the following tables the values of μ and ν are given in the c.g.s. and in the British Engineering systems of units.

The coefficient of viscosity μ of a gas is independent of the pressure and increases with the temperature somewhat less rapidly than the increase of the absolute temperature.

Table 5.

Values of μ for air*.

Temperature	gm./cm. sec.	slug/ft. sec.
0° C.	$1·71 \times 10^{-4}$	$0·358 \times 10^{-6}$
15	1·81	0·378
100	2·21	0·461

When the density ρ is known, the value of the kinematic coefficient of viscosity ν can be deduced at once from this table, since $\nu = \mu/\rho$. The values of ν for air at the standard pressure of 760 mm. of mercury are given in table 6, and in general the value of ν is inversely proportional to the pressure at a given temperature.

* Kaye and Laby, *Physical and Chemical Constants.*

Table 6.

Values of ν for air at standard pressure.

Temperature	cm.²/sec.	ft.²/sec.
0° C.	0·133	$1·43 \times 10^{-4}$
15	0·148	1·59
100	0·234	2·52

Finally, the values of the kinematic coefficient of viscosity ν for water are given in table 7 for comparison with the corresponding values for air.

Table 7.

Values of ν for water.

Temperature	cm.²/sec.	ft.²/sec.
0° C.	0·0179	$1·92 \times 10^{-5}$
5	0·0152	1·63
10	0·0131	1·41
15	0·0115	1·23
20	0·0101	1·08
25	0·0090	0·97

8·23. *Dimensional theory.*

In a perfect fluid the force acting on a body has been expressed in the form
$$F = k\rho V^2 l^2,$$
where ρ is the density of the fluid, V the velocity of the body relative to the fluid and l some typical length of the body. The coefficient k is non-dimensional and depends only on the shape and attitude of the body. This form of expression is the only possible combination of the three fundamental parameters ρ, V and l which will give the dimensions of a force and it can therefore be established without any reference to the flow pattern past the body.

In the case of a viscous fluid there is an additional parameter, the kinematic coefficient of viscosity ν, which has the dimensions of a length multiplied by a velocity. It is now possible to form the non-dimensional function

$$R = \frac{lV}{\nu},$$

and the general expression for the force on a body must be taken to be

$$F = k'\rho V^2 l^2 f\left(\frac{lV}{\nu}\right).$$

This general expression for the force retains the correct dimensions whatever form is given to the function f. In a perfect fluid the viscosity is zero and the function assumes the value

$$f\left(\frac{lV}{\nu}\right) = 1.$$

Also in the special case of laminar flow considered in 8·21, the drag force was shown to be proportional to $\mu V l$ and hence the function f is of the form

$$f\left(\frac{lV}{\nu}\right) = \frac{\nu}{lV}.$$

The usual procedure is to retain the original expression for the force

$$F = k\rho V^2 l^2,$$

and to regard the non-dimensional coefficient k as a function of the non-dimensional parameter $\frac{lV}{\nu}$ or R, which is called the *Reynolds' number* of the flow. If the forces are determined on similar bodies of different size, as for example on an aeroplane and its model in a wind tunnel, the corresponding values of the coefficient k will not have the same values unless the tests are made at the same Reynolds' number. This course is usually impossible, since ν has the same value in both cases while l and V are both smaller in a wind tunnel than in free flight of an aeroplane. It is necessary therefore to investigate the variation of the coefficient k with the Reynolds' number and to establish, if possible, a sound method of extrapolating from the model to the full scale. Variation of the coefficient k with the Reynolds' number is frequently called "scale effect."

Fig. 63 shows the drag coefficient of a long circular cylinder as a function of the Reynolds' number. The drag coefficient is defined by the equation

$$D = C_D \cdot \tfrac{1}{2}\rho V^2 S,$$

where S is the normal projected area, the product of the length and diameter of the cylinder, and the Reynolds' number is taken to be

$$R = \frac{Vd}{\nu},$$

where d is the diameter of the cylinder. This example shows

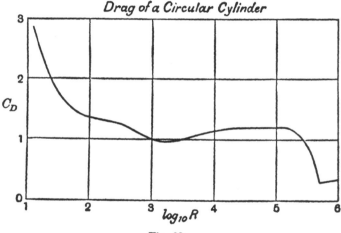

Fig. 63.

that sudden and important changes in the drag coefficient of a body may occur as the Reynolds' number increases. On the other hand, variations of this magnitude are not universal and for many types of body, including aerofoil sections, the drag coefficient is found to tend to a limiting value at an early stage. The variation of the drag coefficient with increasing Reynolds' number is associated with a variation in the flow pattern, and an abrupt change in the drag coefficient implies an abrupt change in the type of flow past the body.

8·24. *Flow in circular pipes.*

The flow along a straight pipe of uniform circular section provides another example of the importance of the Reynolds' number. If r is the radial distance of a cylindrical layer of fluid from the axis of the pipe and if x is the coordinate measured along the axis, the equation of motion for laminar flow will be

$$\frac{d}{dr}\left(2\pi r \mu \frac{du}{dr}\right) = 2\pi r \frac{dp}{dx},$$

or

$$\frac{d}{dr}\left(r \frac{du}{dr}\right) = \frac{r}{\mu}\frac{dp}{dx}.$$

Integrating this equation and inserting the boundary condition of no slip, the velocity is found to be

$$u = -\frac{1}{4\mu}\frac{dp}{dx}(a^2 - r^2),$$

where a is the radius of the pipe. The mean velocity V of the flow is

$$V = \frac{1}{\pi a^2}\int_0^a 2\pi r u\, dr = -\frac{a^2}{8\mu}\frac{dp}{dx},$$

and hence

$$u = 2V\left(1 - \frac{r^2}{a^2}\right).$$

Finally, the drag of length l of the pipe is

$$D = \pi a^2\left(-l\frac{dp}{dx}\right) = 4\mu V\frac{S}{a},$$

where S is the wetted surface $2\pi al$.

The pressure gradient down the pipe in laminar flow is

$$\frac{dp}{dx} = -\frac{8\mu V}{a^2} = -8\left(\frac{\nu}{aV}\right)\frac{\rho V^2}{a},$$

and this result is used to determine the coefficient of viscosity of a fluid from the observed pressure drop along a pipe. It is found by experiment that the laminar flow always establishes itself in a pipe provided the Reynolds' number $\frac{aV}{\nu}$ is less than the critical value 1160, but by suitable precautions to avoid turbulence of the fluid entering the pipe the laminar flow may be continued to far greater values of the Reynolds' number.

For large values of the Reynolds' number the flow is turbulent and the pressure gradient, deduced from a large number of experiments, then obeys the empirical law[*]

$$\frac{dp}{dx} = -0.066 \left(\frac{\nu}{aV}\right)^{\frac{1}{4}} \frac{\rho V^2}{a}.$$

The frictional drag of the surface in turbulent flow is proportional to $V^{1.75}$, contrasting with the case of laminar flow when it is simply proportional to V[†].

This empirical law for the turbulent flow in a circular pipe has been used by Karman[‡] to deduce the law of variation of velocity with distance from the wall of a pipe. In general the surface traction τ, i.e. the force per unit area of the surface, must be of the form

$$\tau = \rho V^2 f(R),$$

where R is the Reynolds' number $\dfrac{aV}{\nu}$ for the flow, and the velocity at distance $y = a\eta$ from the wall of the pipe must be of the form

$$u = VF(\eta, R).$$

Near the wall, however, the velocity u can be expressed also in terms of ρ, ν, y and τ independently of a and V, and by considering the dimensions of these parameters, the form of the velocity must be

$$u = \frac{\nu}{y} \phi \left(\frac{y}{\nu} \sqrt{\frac{\tau}{\rho}}\right).$$

Finally, by equating the two expressions for the velocity u and by eliminating the surface traction τ, an equation is obtained connecting the parameters R and η:

$$\phi(\eta R \sqrt{f}) = \eta RF(\eta, R).$$

In order to obtain a solution of this general equation Karman assumes that the velocity distribution across the pipe is independent of the value of the Reynolds' number R for the range in which Blasius' empirical law is valid. Then

$$\phi(\eta R \sqrt{f}) = \eta RF(\eta),$$

[*] Blasius, *Forschungsarbeiten der V.D.I.* 1913.
[†] See Note 4 of Appendix.
[‡] "Über laminare und turbulente Reibung," *ZAMM*, 1921. For a more general treatment, see also Prandtl, "Bericht über Untersuchungen zur ausgebildeten Turbulenz," *ZAMM*, 1925.

and near the wall it is sufficient to retain only the lowest power of η in the expansions of the functions ϕ and F. Hence

$$(\eta R \sqrt{f})^{1+n} \propto \eta R \eta^n,$$

or

$$f \propto R^{-\frac{2n}{1+n}}.$$

But the empirical law is of the form

$$f \propto R^{-k},$$

and so finally

$$n = \frac{k}{2-k}.$$

The velocity near the wall varies as y^n and the value of n varies with the empirical value of k as follows:

$$k = 1, \tfrac{1}{2}, \tfrac{1}{4}, 0,$$
$$n = 1, \tfrac{1}{3}, \tfrac{1}{7}, 0.$$

In laminar flow $k = 1$ and the velocity varies linearly with the distance from the wall. In the turbulent state Blasius gives $k = \tfrac{1}{4}$ and the velocity varies as the one-seventh power of the distance from the wall. If the value of k decreases further at higher values of the Reynolds' number, then the velocity will rise more rapidly near the wall, and in the limit when the surface traction τ is proportional to ρV^2 simply ($k = 0$), the velocity is uniform across the whole pipe.

The law of variation of velocity with distance from the wall breaks down in the immediate proximity of the wall, since it suggests an infinite value of $\dfrac{\partial u}{\partial y}$ instead of the true finite value $\dfrac{\tau}{\rho \nu}$. This discrepancy is due to the fact that the fluid layer in immediate contact with the wall is always in laminar motion and that the empirical law for the turbulent flow applies only as far as the outer boundary of this laminar layer. Thus the curve $u \propto y^n$ should be accepted down to the point where $\dfrac{\partial u}{\partial y} = \dfrac{\tau}{\rho \nu}$ and should then be continued to the origin by the tangent to the curve.

8·3. *The general equations of motion.*

Hitherto the simple laminar motion of a viscous fluid has been considered, and to discuss the more general types of motion it is necessary to develop the equations of motion of the fluid. In two-dimensional motion the velocity of the fluid at any point is defined by its components u and v parallel to orthogonal axes, and these velocity components must satisfy the equation of continuity (5·1)

$$\frac{\partial u}{\partial x} + \frac{\partial v}{\partial y} = 0.$$

The velocity components u and v define the velocity of the fluid element at the point (x, y). After a small interval of time dt, the fluid element will be at the point $(x + u\,dt, y + v\,dt)$, and the components of the velocity of the fluid element will then be respectively

$$u + \frac{\partial u}{\partial t}\,dt + \frac{\partial u}{\partial x}\,u\,dt + \frac{\partial u}{\partial y}\,v\,dt,$$

$$v + \frac{\partial v}{\partial t}\,dt + \frac{\partial v}{\partial x}\,u\,dt + \frac{\partial v}{\partial y}\,v\,dt.$$

In a perfect fluid the only force acting on the fluid element is the pressure on its boundary which has the components

$$-\frac{\partial p}{\partial x} \quad \text{and} \quad -\frac{\partial p}{\partial y}$$

per unit volume, and hence the equations of motion of the fluid element are

$$\frac{\partial u}{\partial t} + u\,\frac{\partial u}{\partial x} + v\,\frac{\partial u}{\partial y} = -\frac{1}{\rho}\frac{\partial p}{\partial x},$$

$$\frac{\partial v}{\partial t} + u\,\frac{\partial v}{\partial x} + v\,\frac{\partial v}{\partial y} = -\frac{1}{\rho}\frac{\partial p}{\partial y}.$$

In a viscous fluid the element also experiences tangential forces on its boundary, depending on its motion relative to the adjacent fluid elements, and additional terms $\nu\nabla^2 u$ and $\nu\nabla^2 v$ respectively occur on the right-hand side of the equations of motion. The development of these expressions from first principles will be found in a standard text-book on hydrodynamics* and the following discussion is intended only to indicate the physical meaning of the expressions.

If u is the velocity component parallel to the axis of x at

* E.g. Lamb, *Hydrodynamics*, chapter xi.

the point (x, y), then the corresponding velocity component at an adjacent point $(x + \xi, y + \eta)$ may be written as

$$u' = u + \left(\xi \frac{\partial u}{\partial x} + \eta \frac{\partial u}{\partial y} \right) + \frac{1}{2} \left(\xi^2 \frac{\partial^2 u}{\partial x^2} + 2\xi\eta \frac{\partial^2 u}{\partial x \partial y} + \eta^2 \frac{\partial^2 u}{\partial y^2} \right)$$

if higher powers of ξ and η than the second are ignored. The mean value of this velocity component at the four points $(x \pm \xi, y \pm \eta)$ is

$$\bar{u}' = u + \frac{1}{2} \left(\xi^2 \frac{\partial^2 u}{\partial x^2} + \eta^2 \frac{\partial^2 u}{\partial y^2} \right),$$

and for a circular ring of points surrounding the point (x, y) the mean values of ξ^2 and η^2 are equal. Thus

$$\bar{u}' - u \propto \frac{\partial^2 u}{\partial x^2} + \frac{\partial^2 u}{\partial y^2}$$

$$\propto \nabla^2 u.$$

But in the laminar flow considered in 8·2 in defining the viscous force between adjacent fluid elements the velocity component u was a function of y only and the force on unit mass of the fluid was found to be

$$X = \nu \frac{\partial^2 u}{\partial y^2}.$$

This force depends on the relative motion of the adjacent fluid elements and hence in the general case the viscous force per unit mass of the fluid may be expected to be

$$X = \nu \nabla^2 u,$$

with a corresponding expression for the component parallel to the axis of y.

The complete equations of viscous motion in two dimensions are

$$\frac{\partial u}{\partial t} + u \frac{\partial u}{\partial x} + v \frac{\partial u}{\partial y} = -\frac{1}{\rho} \frac{\partial p}{\partial x} + \nu \nabla^2 u,$$

$$\frac{\partial v}{\partial t} + u \frac{\partial v}{\partial x} + v \frac{\partial v}{\partial y} = -\frac{1}{\rho} \frac{\partial p}{\partial y} + \nu \nabla^2 v,$$

and in steady motion the terms $\frac{\partial u}{\partial t}$ and $\frac{\partial v}{\partial t}$ are zero. The solution of these equations for the flow past a body, at whose surface the boundary condition of no slip ($u = v = 0$) must be satisfied, presents almost insuperable difficulties except in a few special cases, and it is necessary to adopt some method of approximation. The conception of a perfect fluid is based on the fact that the viscosity of a fluid is small and that the viscous terms involving ν are negligible in comparison with the dynamic terms involving the square of the velocity. At the other extreme it is possible to consider a slow steady motion of a viscous fluid in which the dynamic terms are negligible in comparison with the viscous terms. The left-hand sides of the equations of motion then disappear, and on eliminating the pressure and expressing the velocity components in terms of the stream function ψ, a single equation is obtained:

$$\nabla^4\psi \equiv \left(\frac{\partial^2}{\partial x^2} + \frac{\partial^2}{\partial y^2}\right)\left(\frac{\partial^2}{\partial x^2} + \frac{\partial^2}{\partial y^2}\right)\psi = 0.$$

Solutions of some problems have been obtained on this basis but they apply only to extremely low velocities. More generally an approximation is required which includes both the dynamic and the viscous terms but reduces the equations to a simpler form.

8·4. *The boundary layer theory.*

Prandtl's approximation to the general equations of viscous motion* is based on the fact that the viscosity of a fluid is small and that it exerts a noticeable effect only where the velocity is changing rapidly from point to point. Now rapid changes of velocity occur only in close proximity to the surface of a body where the velocity rises from zero at the surface to its value in the general stream, and in consequence Prandtl's conception of the problem is that the effect of the viscosity is important only in a narrow boundary layer surrounding the surface of the body and that the viscosity may be ignored in the free fluid outside this layer. In the boundary layer the velocity of the fluid rises rapidly from zero to its value in the free stream, and however small the viscosity may be the viscous force retains its importance in this layer.

* *Verhandl. d. III intern. math. Kongress* (Heidelberg, 1904).

Turning now to the general equations of motion as applied to the boundary layer, the coordinate x will be assumed to be measured along a flat surface and the quantities x, u and p will be finite while y and v will be small of the order ϵ. In the first equation of motion $\frac{\partial^2 u}{\partial x^2}$ is small compared with $\frac{\partial^2 u}{\partial y^2}$ and the equation becomes

$$\frac{\partial u}{\partial t} + u\frac{\partial u}{\partial x} + v\frac{\partial u}{\partial y} = -\frac{1}{\rho}\frac{\partial p}{\partial x} + \nu\frac{\partial^2 u}{\partial y^2},$$

where the last term is of the order $\frac{\nu}{\epsilon^2}$. If ν is small compared with ϵ^2, the last term disappears and the equation becomes that of a perfect fluid. If ν is large compared with ϵ^2, the dynamic terms involving the square of the velocity are negligible and the equation is appropriate to very slow motion. More generally ν must be of the same order as ϵ^2 and the ordinates of the boundary layer are then proportional to $\sqrt{\nu}$.

The second equation of motion leads to the very simple result

$$0 = -\frac{1}{\rho}\frac{\partial p}{\partial y},$$

since all the other terms are small in comparison with this pressure term. This equation shows that the pressure is transmitted normally through the boundary layer without change and hence that the pressure in the boundary layer is a function of the coordinate x only.

The equations governing the flow in the boundary layer are

$$\frac{\partial u}{\partial t} + u\frac{\partial u}{\partial x} + v\frac{\partial u}{\partial y} = -\frac{1}{\rho}\frac{dp}{dx} + \nu\frac{\partial^2 u}{\partial y^2},$$

$$\frac{\partial u}{\partial x} + \frac{\partial v}{\partial y} = 0.$$

These equations have been developed for the flow along a flat surface, but the identical form can be obtained more generally for a curved surface if the coordinate x is measured along the surface and the coordinate y normal to it.

8·41. *Drag of a flat plate.*

The boundary layer theory has been applied by H. Blasius[*] to the determination of the laminar flow along a flat plate and of the resulting frictional drag. Measuring the coordinate x along the plate from the leading edge, the thickness of the

Fig. 64.

boundary layer is shown to be proportional to $\sqrt{\dfrac{\nu x}{V}}$ and the frictional drag of both surfaces of a plate of length c to be

$$D = 1\cdot328\,\sqrt{\frac{\nu}{cV}}\,c\rho V^2$$

per unit breadth. Thus the drag is proportional to $V^{1\cdot5}$ and the drag coefficient of the flat plate, regarded as an aerofoil, is

$$C_D = 2\cdot656\,\sqrt{\frac{\nu}{cV}}.$$

The thickness of the boundary layer cannot be determined exactly, as the velocity u in the boundary layer tends asymptotically to the velocity V of the free stream, but if the outer surface of the layer is defined by the condition that u has risen to a value differing from V by two per cent., the thickness of the boundary layer may be taken to be

$$\delta = 4\cdot5\,\sqrt{\frac{\nu x}{V}}.$$

The value $cV = 10^6\,\nu$ marks the division between the model and "full scale" range for an aerofoil and with this value the maximum thickness of the boundary layer is $0\cdot0045c$.

Blasius' solution corresponds to laminar flow along the plate and will represent the actual flow at low values of the Reynolds' number only. Karman[†] has obtained a solution

[*] "Grenzschichten in Flüssigkeiten mit kleiner Reibung," *Zeitschrift f. Math. u. Phys.* 1908.

[†] "Über laminare und turbulente Reibung," *ZAMM*, 1921.

for the turbulent flow along a flat plate by analysing Blasius'
empirical law for the turbulent flow in a pipe (8·24) and has
obtained the drag coefficient*

$$C_D = 0·144\left(\frac{\nu}{cV}\right)^{0·2}.$$

Thus the drag is proportional to $V^{1·8}$ and the thickness of
the boundary layer to $x^{0·8}$. Karman's result may be compared
with the experimental determination†

$$C_D = 0·075\left(\frac{\nu}{cV}\right)^{0·15},$$

and for the range of the experiments ($R = 3 \times 10^5$ to 7×10^6)
the numerical values given in table 8 show good agreement.
The numerical values given by Blasius' formula are added
for comparison and show that the change from laminar to
turbulent flow causes an increase of drag.

Table 8.

Frictional drag coefficient of a flat plate.

$R =$	3×10^5	10^6	7×10^6
Experimental	0·0114	0·0094	0·0070
Karman	0·0116	0·0090	0·0061
Blasius	0·0048	0·0026	0·0010

8·42. The boundary layer theory can also be used to
explain the phenomenon of
the flow breaking away from
the surface of a body to
form an eddying wake. In a
perfect fluid the streams passing above and below a body
unite behind the body and
there is a stagnation point S
on the surface. While passing
from A to S the velocity of

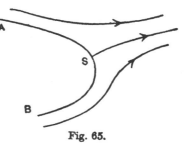

Fig. 65.

the fluid decreases and the pressure increases, and the fluid

* See Note 5 of Appendix.
† *Ergebnisse der aerodynamischen Versuchsanstalt zu Göttingen*, I, 1921.

elements lose kinetic energy in forcing their way along the surface against the increasing pressure. When viscous forces also occur in the boundary layer adjacent to the surface, the fluid elements will lose energy more rapidly and will be brought to rest before reaching the point S, and a reverse flow will set in from S towards A as indicated in fig. 66. The same process will occur on the lower surface of the body and thus two surfaces of dis-

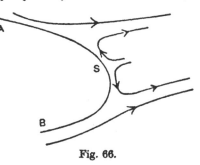

Fig. 66.

continuity will arise, as assumed in the Helmholtz-Kirchhoff theory. These surfaces of discontinuity are unstable and lead to the development of a Karman vortex street behind the body.

The condition for the flow to break away from the surface is associated with an increase of pressure along the surface, and for a given pressure distribution along the surface it is possible to calculate the point at which the flow breaks away by means of the equations of the boundary layer. Unfortunately it is not sufficiently accurate to assume the pressure distribution given by the perfect fluid solution, and even when the points of origin of the surfaces of discontinuity have been determined, a further advance of the theory is required to determine the strength and breadth of the resulting vortex street. The problem of the form drag of a body therefore remains to be solved, although the theory indicates the formation of the surfaces of discontinuity and the nature of the final vortex street*.

* See Note 6 of Appendix.

CHAPTER IX

THE BASIS OF AEROFOIL THEORY

9·1. The theory of the lift force given by an aerofoil in two-dimensional motion has been developed by considering the flow of a perfect fluid governed by Joukowski's hypothesis that the flow leaves the trailing edge of the aerofoil smoothly. It is necessary now to examine the fundamental basis of this theory and the extent to which the assumed motion represents the actual conditions which occur with a viscous fluid.

All real fluids possess the property of viscosity and the conception of a perfect fluid should be such that it represents the limiting condition of a fluid whose viscosity has become indefinitely small. Now it is well known that the limit of a function $f(x)$ as x tends to zero is not necessarily equal to the value of the function when x is equal to zero, and hence, to obtain the true conception of a perfect fluid, it is not sufficient to assume simply that the coefficient of viscosity is zero. The viscosity must be retained in the equations of motion and the flow for a perfect fluid must be obtained by making the viscosity indefinitely small.

9·2. Slip on the boundary.

The first point to be considered is the motion of the fluid at the surface of a body. In a viscous fluid the relative velocity at the surface of a body is zero and the body is surrounded by a narrow boundary layer in which the velocity rises rapidly from zero to a finite value. The thickness of this boundary layer, which is essentially a region of vorticity, is proportional to $\sqrt{\nu}$ and tends to zero with the viscosity. Thus in the limit the boundary layer becomes a vortex sheet surrounding the surface of the body and the vortices of this sheet act as the roller bearings between the surface of the body and the general mass of the fluid. The assumption of a perfect fluid with a vortex sheet surrounding the surface of the body therefore represents the limiting condition of a viscous fluid when the viscosity tends to zero, and the existence of the vortex sheet implies that the perfect fluid solution need not satisfy the condition of zero slip at the boundary. If q is the velocity at the surface in the perfect

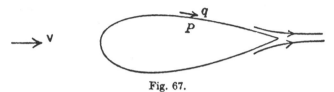

Fig. 67.

fluid solution, then the strength of the vorticity of the vortex sheet will be q per unit length. If this vortex sheet is assumed to surround the surface of the body, the condition of zero slip at the boundary is satisfied, but the velocity rises from zero to the value q in passing through the indefinitely thin vortex sheet, and the conditions external to the boundary layer are identically the same as if the vortex sheet were ignored and the condition of zero slip at the boundary were abandoned. The sum of the strengths of the vortices composing the vortex sheet is equal to the magnitude of the circulation round the body in the perfect fluid solution.

The boundary layer transmits the pressure through itself normally without alteration and hence the actual pressure distribution on the surface of the body will be identical with that obtained from the perfect fluid solution by means of Bernoulli's equation*.

In proceeding to the limit of zero viscosity it is necessary to retain the actual type of flow which occurs with a viscous fluid. Thus in the case of a circular cylinder the flow breaks away from the surface in two vortex sheets which develop into a Karman vortex street and this type of flow must be retained in the limiting case. The type of flow considered in 3·6, where the flow passes smoothly to the rear of the cylinder, is clearly inadmissible and does not represent even an approximation to the actual flow except possibly near the nose of the cylinder (cf. fig. 14). The position on the cylinder

* See Note 7 of Appendix.

at which the flow breaks away from the surface and the nature of the resulting vortex street will depend on the magnitude of the viscosity*,but it can easily be seen that this characteristic of the flow cannot disappear as the viscosity tends to zero. Referring to fig. 67, where the surface of the body is supposed to be surrounded by a vortex sheet, the vortex element at P has the velocity $\frac{1}{2}q$ along the surface (cf. 4·35) and hence fluid elements in vortex motion are continually passing along the surface of the body from front to rear. These vortex elements must leave the body eventually and pass down stream in a vortex wake which is the Karman vortex street of the body. The breadth and strength of this vortex street will depend on the shape of the body, but in all cases it is necessary to presume the existence of a vortex wake of this type.

9·3. *Joukowski's hypothesis.*

The motion of a perfect fluid past an aerofoil can be determined with any arbitrary circulation of the flow round the aerofoil, but in the development of the theory the circulation round an aerofoil with a sharp trailing edge was determined by means of Joukowski's hypothesis that the flow must leave the trailing edge smoothly. With any other value of the circulation the velocity of the fluid would become infinite at the trailing edge and the viscous force at this point could not be neglected even when the viscosity became indefinitely small, for however small a value were assigned to ν it would always be possible to find a region close to the trailing edge of the aerofoil where the product of ν and the rate of change of velocity $\frac{\partial q}{\partial n}$ was large. Hence it follows that the only perfect fluid solution which can be regarded as the limit of the true viscous fluid solution is that which avoids an infinite velocity at the trailing edge, and this solution is defined by Joukowski's hypothesis.

The conception that the fluid velocity must be finite at all points can be applied more generally as a criterion of the validity of any perfect fluid solution. Thus the flow past a

* See Note 8 of Appendix.

line shown in fig. 43 is clearly impossible as the limit of a real viscous fluid solution as the viscosity tends to zero, and the actual motion must be of the type where the flow breaks away from the surface at the ends of the line (fig. 56).

The magnitude of the circulation round an aerofoil determined by Joukowski's hypothesis is not quite accurate, since it ignores the influence of the vortex wake which is formed behind the aerofoil. The flow on the under surface of the aerofoil, along which the pressure decreases towards the trailing edge, will pass continuously to the trailing edge and leave it smoothly, but the flow on the upper surface, along which the pressure increases towards the trailing edge, will break away from the surface before reaching the trailing edge to form the upper boundary of the vortex wake*. In consequence the true circulation will be slightly less than that determined by Joukowski's hypothesis. It appears, however, that aerofoils of good shape at small angles of incidence have an extremely small form drag (see 9·5) and the vortex wake must be too weak and too narrow to exert a noticeable effect on the circulation. At large angles of incidence the vortex wake is more important since the flow breaks away from the upper surface of the aerofoil at an earlier stage. Joukowski's hypothesis then breaks down completely, the lift ceases to rise with the angle of incidence and the aerofoil reaches its critical angle. In this region the aerofoil theory of chapter VII is no longer valid, the aerofoil must be regarded as a bluff body and the most important feature of the flow is the vortex wake rather than the circulation.

In the ordinary working range of an aerofoil Joukowski's hypothesis can be used to determine the magnitude of the circulation with good accuracy, and this determination is independent of the exact value of the viscosity, which has merely been assumed to be very small. Hence no appreciable scale effect on the lift of an aerofoil is to be anticipated in this range. On approaching the critical angle, however, the flow breaks away from the upper surface of the aerofoil to form a broad vortex wake, and there may be an important

* See Note 9 of Appendix.

scale effect on the lift of the aerofoil since the nature of the vortex wake will depend on the Reynolds' number of the flow.

9·4. *Origin of the circulation.*

The process by which the circulation round an aerofoil develops as the aerofoil starts from rest presents certain theoretical difficulties, since this process would be impossible in a perfect fluid, and it is again necessary to consider the limiting condition as the viscosity tends to zero. At extremely low speeds when the aerofoil is starting from rest the

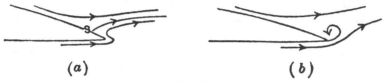

(a)   (b)

Fig. 68.

flow near the trailing edge will be of the type shown in fig. 68 (a), with a stagnation point S on the upper surface at some small distance from the trailing edge. As the forward velocity of the aerofoil increases, the stream lines along the under surface are no longer able to turn round the trailing edge owing to the large viscous forces brought into action by the high velocity gradient, the flow breaks away from the trailing edge, and a vortex is formed between the trailing edge and the old stagnation point S as shown in fig. 68 (b). When this vortex has developed to a certain stage, it breaks away from the aerofoil and passes down stream in the vortex wake. Now the circulation round any large contour $ABCD$ (fig. 69) which surrounded the aerofoil initially was and

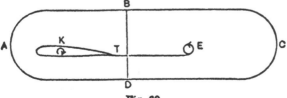

Fig. 69.

must remain zero, and as this contour includes the vortex E there must be a circulation K round the aerofoil which is exactly equal and opposite to the circulation round the vortex E. In the course of time the vortex E passes far down stream where it can no longer influence the flow round the aerofoil, and the aerofoil is then in steady motion with a circulation of the flow round it.

The existence of the vortex E in the early stages of the motion can be verified experimentally in a very simple manner by dipping a flat plate into water and moving it briskly in a direction inclined at a small angle to its surface. If the motion develops gradually instead of impulsively, a succession of vortices will be shed from the trailing edge of the aerofoil, but the previous argument remains valid and the resulting circulation round the aerofoil is equal in magnitude to the sum of the strengths of the vortices which have left the aerofoil*.

The general magnitude of the circulation round an aerofoil is determined by the strength of the vortices which were shed in the initial stages of the motion or at any time when the speed or attitude was changed, but in addition the magnitude of the circulation is subject to a small fluctuation. The vorticity of the boundary layer passes down stream in a vortex wake which develops into a Karman vortex street and to maintain this system vortices of opposite sign are shed alternately from the upper and lower surfaces of the aerofoil†. Since the sum of the circulation round the aerofoil and of the strengths of all the vortices of the wake must be zero, it follows that the circulation round the aerofoil will oscillate between the limits $K \pm \frac{1}{2}k$, where K is the mean circulation and k is the strength of the vortex street. For a good aerofoil section at a small angle of incidence the vortex wake is narrow and weak, and the circulation round the aerofoil is sensibly constant, but as the attitude of the aerofoil approaches and passes its critical angle the oscillation in the magnitude of the circulation may become an important fraction of the mean circulation.

* See Note 10 of Appendix.   † See Note 3 of Appendix.

9·5. *The drag of an aerofoil.*

In developing the theory of the lift of an aerofoil, the drag was neglected completely, and this method is justified solely by the fact that the drag is so small a fraction of the lift that the modification of the flow necessary to explain the drag does not exert a noticeable influence on the characteristics of the flow which determine the lift. The method will clearly break down near the critical angle where the drag increases rapidly owing to the development of a strong vortex wake, and also near the angle of no lift where the lift and drag are of the same order of magnitude. Since the drag depends on the viscosity and varies with the Reynolds' number of the flow, scale effect on the lift may be anticipated in the neighbourhood both of no lift and of the critical angle.

The drag of an aerofoil in two-dimensional motion is called the *profile drag*, since it depends essentially on the shape of the aerofoil section or profile. The profile drag may be considered in two parts, the form drag associated with the vortex street behind the aerofoil and the frictional drag on the surface of the aerofoil[*]. A measurement of the pressure distribution over the surface of an aerofoil can be used to determine the lift and the form drag, but the frictional drag cannot be determined by this method.

The profile drag coefficient of a good aerofoil section is extremely low, and the following table gives the values of the minimum profile drag coefficients of a few aerofoil sections for the value of the Reynolds' number $R = 2·5 \times 10^5$, at which the frictional drag coefficient of a flat plate is 0·0116. In the case of a thin symmetrical section, Göttingen 443, a profile drag coefficient as low as 0·0054 has been obtained at $R = 4 \times 10^5$, and this value is only half the frictional drag

Table 9.

Minimum profile drag coefficients.

RAF 15	0·0116
RAF 25	0·0080
RAF 30	0·0112

[*] See Note 3 of Appendix.

coefficient of a flat plate at the same value of the Reynolds' number*.

These experimental values justify the assumption, made in developing the theory of an aerofoil, that the drag is negligibly small compared with the lift over the ordinary working range of incidence. It also appears that the profile drag of an aerofoil section may be less than the frictional drag of a flat plate of the same chord. The form drag of the aerofoil must therefore be extremely small and the existence of the vortex wake can be ignored with safety in determining the magnitude of the circulation by Joukowski's hypothesis.

* See Note 11 of Appendix.

CHAPTER X

THE AEROFOIL IN THREE DIMENSIONS

10·1. *Circulation and vorticity.*

The definition of the circulation round a closed curve in two dimensions (see 4·1) as the integral of the tangential component of the velocity round the circumference of the curve can be extended at once to the more general case of motion in three dimensions by removing the restriction that the curve must lie in a single plane. Also by dividing any surface bounded by this curve into a network by a series of intersecting lines it can be shown that the circulation round the curve is equal to the sum of the circulations round the elementary areas formed by the network.

The vorticity of a fluid element in two-dimensional motion was defined (see 4·3) as twice the angular velocity of the element. This definition is retained in the more general case of three-dimensional motion but the axis of rotation of the fluid element may now point in any direction. By following the direction of the axis of rotation of successive fluid elements it is possible to construct a curved line whose direction coincides at every point of its length with the axis of rotation of the corresponding fluid element. Such a line is called a *vortex line*.

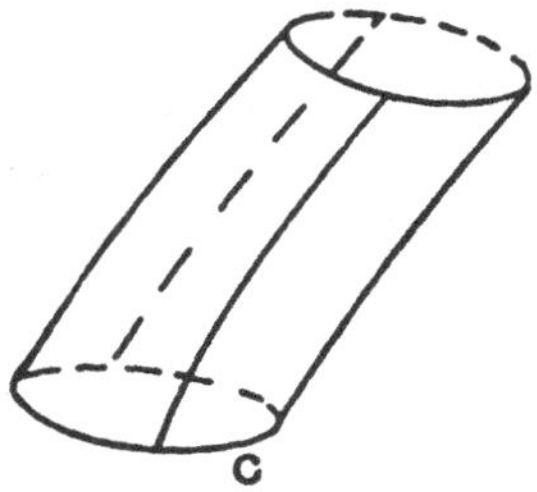

The vortex lines which pass through the points of the circumference of a small closed curve C will form the surface of a *vortex tube*, of which the curve C is a cross section. If 2ω is the vorticity at this section of the vortex tube and if the section is taken at right angles to the axis of the tube, the circulation K round the curve C

Fig. 70.

will be equal to $2\omega S$, the product of the vorticity and the area of the cross section. If the section is taken so that its normal is inclined at angle θ to the axis of the tube, the area of the section will be increased to $S \sec \theta$, but the component of the angular velocity about the normal to the section will be reduced to $\omega \cos \theta$, and the circulation which is equal to twice the product of these two quantities is unaltered. Also the circulation round any small curve which lies on the surface of the vortex tube will be zero, since the component of the angular velocity normal to the surface of the tube is essentially zero.

If a curve of the type shown in fig. 71 is drawn on the surface of a vortex tube, the circulation round the curve will be zero. Now if $F(AB)$ denotes the flow along the curve AB, this result can be expressed in the form

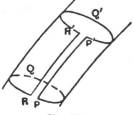

Fig. 71.

$$F(PQR) + F(RR') + F(R'Q'P') + F(P'P) = 0,$$

and when PP' coincides with RR' this equation becomes

$$F(PQR) = F(P'Q'R'),$$

showing that the circulation has the same value for all curves embracing the vortex tube. The value of this circulation K is called the strength of the vortex tube.

10·11. The conception of a *line vortex* is derived from that of a vortex tube by making the area of cross section of the tube tend to zero while the strength K remains unaltered. The line vortex in three-dimensional motion corresponds to the point vortex in two-dimensional motion, but whereas the latter represents a straight line of infinite length normal to the plane in which the two-dimensional motion occurs, a line vortex may in general be a curve of any shape. The circulation round any closed curve C is equal to the sum of the strengths of the line vortices which cut any surface bounded by this curve, and from this fact it follows that a line vortex cannot come to an end in the fluid. It must form a closed curve or have its ends on a solid boundary. A line vortex is exactly analogous to a wire carrying an electric current, the strength of the line vortex corresponds to the strength of the electric current, and the induced velocity at any point of the fluid corresponds to the magnetic force due to the electric current.

10·12. The induced velocity of an element of a line vortex at a point P is determined by the equation*

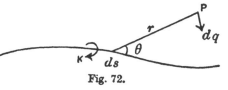

Fig. 72.

$$dq = \frac{K\,ds}{4\pi r^2}\sin\theta,$$

where K is the strength and ds is an element of length of the line vortex, r is the distance of the point P from the element, and θ is the angle between the direction of the element and the line joining the element to the point P. The velocity dq is normal to the plane containing r and ds, and its sense is the same as that of the circulation K about the line vortex.

An element ds of a line vortex cannot exist independently and the formula should be used only for integrating the effect of a complete line vortex. Frequently, however, a line vortex may be built up of a number of straight lines and it is useful therefore to determine the induced velocity of a straight line vortex of finite length AB. If PN, the normal

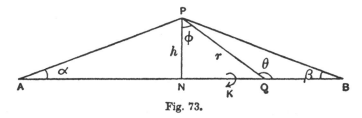

Fig. 73.

from any point P to the line AB, is of length h, and if Q is any point of the line vortex, the induced velocity at the point P due to the element ds at Q is

$$dq = \frac{K\,ds}{4\pi r^2}\sin\theta = \frac{Kh\,ds}{4\pi r^3},$$

* Cf. Lamb, *Hydrodynamics*, § 149.

and this velocity is normal to the plane PAB. Now if ϕ is the angle NPQ, the element of length ds may be expressed as

$$ds = d\,(h \tan \phi) = h \sec^2 \phi\, d\phi,$$

and hence

$$dq = \frac{K}{4\pi h} \cos \phi\, d\phi.$$

The total induced velocity of the line vortex AB is obtained by integration from $\phi = -\left(\dfrac{\pi}{2} - \alpha\right)$ to $\phi = \left(\dfrac{\pi}{2} - \beta\right)$, where α and β are the angles PAB and PBA respectively. Thus finally

$$q = \frac{K}{4\pi h}\,(\cos \alpha + \cos \beta).$$

If the line AB is of infinite length, this result reduces to

$$q = \frac{K}{2\pi h},$$

which agrees with the formula for the induced velocity of a point vortex in two-dimensional motion.

It is also important to note that the induced velocity of a line, which starts at the point N and extends to infinity in one direction only, is

$$q = \frac{K}{4\pi h},$$

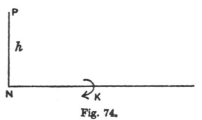

Fig. 74.

as this result is used repeatedly in the development of aerofoil theory.

10·2. *The vortex system of an aerofoil.*

In dealing with the problem of an aerofoil of finite span in three-dimensional motion the assumptions will be made that the chord of the aerofoil is small compared with the span, that the span may be regarded as a straight line at right angles to the direction of motion, and that the aerofoil is symmetrical laterally about the mid-point of its span. Apart from these restrictions the chord, angle of incidence and shape of the aerofoil section may vary in any manner across the span of the aerofoil.

If the aerofoil experiences a lift force there must be a circulation of the flow round the aerofoil sections and therefore in effect there is a line vortex or a set of line vortices running along the span of the aerofoil. These line vortices, which move with the aerofoil, are called the *bound vortices* of the aerofoil, and are formed by the boundary layer or vortex sheet which surrounds the surface of the aerofoil. In accordance with the general theory of vortex motion, these line vortices cannot end at the tips of the aerofoil but must continue in the fluid as free line vortices. Also any element of the fluid, which is set in vortex motion by coming into contact with the bound vortex system of the aerofoil, will

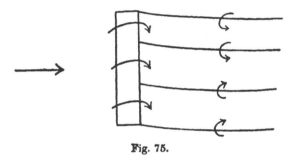

Fig. 75.

pass down stream with the general mass of the fluid, and free line vortices will therefore start at the surface of the aerofoil and pass down stream along the stream lines of the flow as indicated in fig. 75. These line vortices are called the *trailing vortices* of the aerofoil.

The vortex system is completed far behind the aerofoil by a transverse vortex parallel to the span of the aerofoil, which is the vortex shed from the trailing edge at the commencement of the motion (cf. 9·4). For all practical purposes, however, the trailing vortices may be assumed to extend down stream indefinitely.

10·21. The simplest type of vortex system occurs when the circulation round the aerofoil sections has a constant value K across the span of the aerofoil. The bound vortex system can then be conceived as a single line vortex of strength K, and the trailing vortices will be two line vortices of the same strength which spring from the tips of the aerofoil and pass down stream in the direction of the stream lines. These line vortices will be curved owing to the variation in the downward component of the velocity at different distances behind the aerofoil, but for most purposes

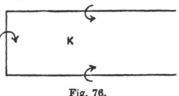

Fig. 76.

it is sufficiently accurate to assume that they are straight lines parallel to the direction of motion. In this way the simple conception of a "horseshoe" vortex system, shown in fig. 76, is obtained.

The actual vortex system of an aerofoil is more complicated than this simple system owing to the fact that the

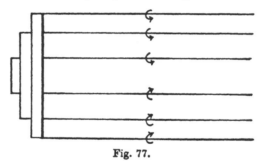

Fig. 77.

circulation is not constant across the span of an aerofoil but generally has a maximum value at the centre and decreases to zero at the tips. Any distribution of circulation across the span can be built up by superimposing a number of the simple "horseshoe" systems, and hence it appears that the free vortex system of an aerofoil will in general consist of a sheet of trailing vortices, springing from the trailing edge of the aerofoil.

10·22. The origin of the trailing vortex system may be considered also from a slightly different point of view. If the distribution of lift across the span of an aerofoil has a maximum value at the centre, there will be a large increase

of pressure below the centre of the aerofoil and a large reduction of pressure above it, and these pressure differences will
decrease towards the tips of the
aerofoil (fig. 78). As a consequence
of this pressure distribution, the
stream lines passing above the aerofoil tend to flow inwards towards

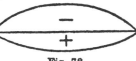

Fig. 78.

the centre and those passing below the aerofoil tend to flow
outwards. As these streams leave the trailing edge of the
aerofoil they form a surface of discontinuity (fig. 79) and the
trailing vortices of the aerofoil represent the vorticity of this
surface of discontinuity.

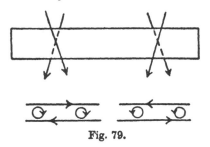

Fig. 79.

10·23. The surface of discontinuity represented by the
sheet of trailing vortices is unstable and will roll up into a
pair of vortex tubes which extend down stream at a distance

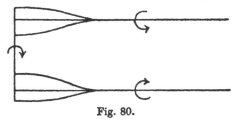

Fig. 80.

apart rather less than the span of the aerofoil (see 12·4). The
trailing vortex system is therefore of the type* shown in
fig. 80. The influence of the trailing vortex system near the

* This type of vortex system was predicted by Lanchester, *Aerodynamics*, 1908.

aerofoil is represented with sufficient accuracy by assuming that the individual line vortices, which spring from the trailing edge of the aerofoil, extend down stream as straight lines. For points of the wake it is more accurate to assume a vortex system of the "horseshoe" type with span rather less than that of the aerofoil, and for points distant from the aerofoil and its wake either representation may be used with equal accuracy.

10·3. *The induced velocity.*

The flow at any section of the aerofoil differs from the flow which would occur round the section in two-dimensional motion owing to the influence of the trailing vortex system. The induced velocity of this vortex system is normal to the span of the aerofoil and to the direction of motion, and is directed

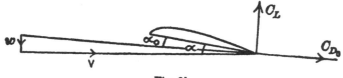

Fig. 81.

downwards in general. The normal induced velocity at a point of the aerofoil will be denoted by w and will be assumed to be small in comparison with the velocity V of the general stream of the fluid. The effect of the induced velocity is then equivalent to a reduction of the angle of incidence of the aerofoil section by the small angle $\frac{w}{V}$ (fig. 81), and if α is the geometrical angle of incidence of the aerofoil section, the effective angle of incidence will be

$$\alpha_0 = \alpha - \frac{w}{V}.$$

More accurately the induced velocity should be regarded as variable along the chord of the aerofoil section, resulting in a change of effective camber of the aerofoil section, but the theory of an aerofoil of finite span can be developed with sufficient accuracy by assuming the chord of the aerofoil section to be small and by assuming a constant value of the induced velocity along the chord. The component of the velocity parallel to the span of the aerofoil is also neglected in developing the theory, since this component is small and unimportant, except possibly at the tips of the aerofoil.

The aerofoil section behaves exactly the same as if it formed part of an aerofoil of infinite span at an angle of incidence α_0, and gives the lift coefficient C_L and the profile drag coefficient C_{D_0} corresponding to two-dimensional motion at this angle of incidence. The lift force is, however, inclined backwards at the small angle $\frac{w}{V}$ (fig. 81) and therefore gives a component in the direction of the drag force. This component is called the *induced drag*, since it is caused by the induced velocity of the trailing vortices. The induced drag coefficient of the aerofoil section is

$$C_{D_1} = \frac{w}{V} C_L,$$

and the total drag coefficient of the aerofoil section as part of the monoplane aerofoil is

$$C_D = C_{D_0} + \frac{w}{V} C_L.$$

The work done on the fluid by the induced drag of the aerofoil appears as the kinetic energy of the trailing vortex system, which increases in length as the motion proceeds.

Since the aerofoil section behaves exactly as in two dimensional motion there is no change in the moment coefficient or in the position of the centre of pressure at any definite value of the lift coefficient.

The characteristics of a monoplane aerofoil are determined by finding the normal induced velocity w and the effective angle of incidence α_0 at each point of the span, by calculating the corresponding elementary lift and drag forces, and by integrating across the span of the aerofoil. The first stage of the calculation of the characteristics of a finite monoplane aerofoil is therefore the determination of the normal induced velocity at a point of the aerofoil in terms of the strength of the trailing vortices.

10·31. The simplest type of trailing vortex system occurs when the circulation has a constant value K across the span of the aerofoil. This case of uniform loading and the simple "horseshoe" vortex system do not represent the true conditions for any actual aerofoil, and the system is considered here only as a simple example of the calculation of the normal induced velocity.

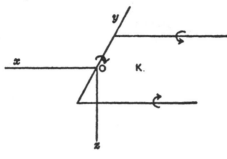

Fig. 82.

The lift of the aerofoil of area S and semi-span s can be expressed in the alternative forms

$$L = C_L \cdot \tfrac{1}{2} \rho V^2 S = 2s\rho V K,$$

and hence
$$K = \frac{S}{4s} C_L V = \tfrac{1}{2} C_L c V,$$

where c is the mean chord of the aerofoil. The trailing vortex system consists simply of two trailing vortices of strength K springing from the tips of the aerofoil, and using the standard system of coordinate axes as shown in fig. 82 with origin at the centre of the aerofoil, the normal induced velocity at a point of the aerofoil is

$$w = \frac{K}{4\pi (s-y)} + \frac{K}{4\pi (s+y)} = \frac{K}{2\pi} \frac{s}{s^2 - y^2},$$

or
$$\frac{w}{V} = \frac{C_L}{2\pi A} \frac{s^2}{s^2 - y^2},$$

where A is the aspect ratio $\dfrac{2s}{c}$ of the aerofoil. The induced velocity w in this case has a minimum value at the centre and rises to infinity at the tips of the aerofoil, and it is on account of this excessive velocity that the simple "horseshoe" vortex system cannot represent the true conditions for any aerofoil.

10·32. In general the circulation K round an aerofoil will vary across the span, being symmetrical about the centre and falling to zero at the tips. Between the points y and

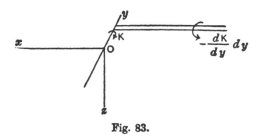

Fig. 83.

$(y + dy)$ of the span of the aerofoil the circulation falls by the amount $-\dfrac{dK}{dy}\,dy$ and hence a trailing vortex of this strength springs from the element dy of the span as shown in fig. 83. There is therefore a sheet of trailing vortices extending across the span of the aerofoil and the normal induced velocity at any point of the span must be obtained as the sum of the effects of all the trailing vortices of this sheet. The normal induced velocity at the point y_1 of the aerofoil is

$$w\,(y_1) = \int_{-s}^{s} \frac{-\dfrac{dK}{dy}\,dy}{4\pi\,(y - y_1)}$$

$$= \frac{1}{4\pi} \int_{-s}^{s} \frac{\dfrac{dK}{dy}\,dy}{y_1 - y}.$$

The evaluation of the integral needs some special care since the integrand becomes infinite at the point $y = y_1$. The value of the integral must be determined by integrating from $-s$ to $y_1 - \epsilon$ and from $y_1 + \epsilon$ to s, and by proceeding to the limit as ϵ tends to zero.

CHAPTER XI

THE MONOPLANE AEROFOIL

11·1. *The fundamental equations.*

If K is the circulation round any section of an aerofoil, the normal induced velocity at a point y_1 of the span is determined by the equation

$$w(y_1) = \frac{1}{4\pi} \int_{-s}^{s} \frac{\frac{dK}{dy}\,dy}{y_1 - y},$$

and the typical aerofoil section experiences the lift force corresponding to two dimensional motion at the effective angle of incidence

$$\alpha_0 = \alpha - \frac{w}{V}.$$

The direction of the line of action of this force component is rotated backwards through the small angle $\frac{w}{V}$ (cf. fig. 81) and hence the drag of the aerofoil section is the profile drag increased by the induced drag, the product of $\frac{w}{V}$ and the lift of the aerofoil section.

Now if the angles of incidence α and α_0 are measured from the attitude of no lift, the lift coefficient of the aerofoil section under these conditions will be

$$C_L = a_0 \alpha_0,$$

where a_0 is the slope of the curve of lift coefficient against angle of incidence for the aerofoil section in two-dimensional motion. Also the circulation K round the aerofoil section will be

$$K = \tfrac{1}{2} C_L c V = \tfrac{1}{2} a_0 c\,(V\alpha - w),$$

and this is a second equation connecting the circulation K and the normal induced velocity w. By means of these two equations it is possible to determine the circulation and the normal induced velocity for any aerofoil in terms of the chord and angle of incidence of the aerofoil sections, which may of course vary across the span of the aerofoil.

Strictly the quantity a_0 should be regarded as a variable depending on the shape of the aerofoil section, but the theory of an aerofoil in two-dimensional motion has shown that a_0 is approximately equal to 2π for all practical aerofoil sections, and hence the variability of a_0 may be neglected without any appreciable loss of accuracy. Nevertheless, since an aerofoil section may fail to realise the theoretical value $a_0 = 2\pi$, the theory of the aerofoil of finite span will be developed in terms of a_0 as the slope of the curve of lift coefficient against angle of incidence in two-dimensional motion, and the theoretical value $a_0 = 2\pi$ will be used only in numerical illustrations of the general formulae.

When the circulation K and the normal induced velocity w of any monoplane aerofoil have been determined, the lift and induced drag are obtained by evaluating the integrals

$$L = \int_{-s}^{s} \rho V K \, dy,$$

$$D_1 = \int_{-s}^{s} \rho w K \, dy.$$

11·2. Method of solution.

A convenient method of attacking the problem of any monoplane aerofoil is to replace the coordinate y, measured to starboard along the span of the aerofoil from its centre, by the angle θ defined by the equation

$$y = - s \cos \theta,$$

so that θ varies from 0 to π across the span of the aerofoil from port to starboard. The circulation K, which is a function of y, may then be expressed as the Fourier series

$$K = 4sV \sum_{n-1}^{\infty} A_n \sin n\theta,$$

and the values of the coefficients A_n must be determined in accordance with the two fundamental equations connecting K and w. The series chosen for the circulation K satisfies the condition that the circulation falls to zero at the tips of the aerofoil, and since the aerofoil is symmetrical about its midpoint odd integral values only of n will occur in the series.

The normal induced velocity at the point y_1 or θ_1 of the aerofoil now becomes

$$w(\theta_1) = \frac{V}{\pi} \int_0^\pi \frac{\Sigma n A_n \cos n\theta}{\cos\theta - \cos\theta_1}\, d\theta$$

$$= V\Sigma n A_n \frac{\sin n\theta_1}{\sin\theta_1},$$

since *
$$\int_0^\pi \frac{\cos n\theta\, d\theta}{\cos\theta - \cos\phi} = \pi \frac{\sin n\phi}{\sin\phi}.$$

Thus at the general point θ of the aerofoil

$$w\sin\theta = V\Sigma n A_n \sin n\theta.$$

The second equation connecting the circulation and the normal induced velocity becomes

$$4sV\Sigma A_n \sin n\theta = \tfrac{1}{2}a_0 c V\left\{\alpha - \frac{\Sigma n A_n \sin n\theta}{\sin\theta}\right\},$$

or
$$\Sigma A_n \sin n\theta\, (n\mu + \sin\theta) = \mu\alpha\sin\theta,$$

where
$$\mu = \frac{a_0 c}{8s}.$$

This is the fundamental equation for determining the values of the coefficients A_n for any monoplane aerofoil. The equation must be satisfied at all points of the aerofoil, but since the aerofoil is symmetrical about its mid-point it is sufficient to consider values of θ between 0 and $\frac{\pi}{2}$. The parameter μ, which is proportional to the chord c, and the angle of incidence α must be regarded as functions of θ in the most general case.

11·21. *Lift and induced drag.*

The lift and induced drag of a monoplane aerofoil are determined very simply in terms of the coefficients A_n of the series for the circulation. The lift of the aerofoil is

$$L = \int_{-s}^s \rho V K\, dy$$

$$= \int_0^\pi 4s^2\rho V^2\,(\Sigma A_n \sin n\theta)\sin\theta\, d\theta$$

$$= 2\pi s^2\rho V^2 A_1,$$

* See page 92.

or, in terms of the lift coefficient,

$$A_1 = \frac{S}{4\pi s^2} C_L.$$

It appears that the lift of the aerofoil is determined by the value of the coefficient A_1 and that the other coefficients of the series for the circulation modify the shape of the load grading curve across the span of the aerofoil without altering the total lift.

The expression $\frac{S}{4\pi s^2}$ which occurs in the equation connecting A_1 and C_L can be expressed in an alternative form. The mean chord of an aerofoil is defined as the area divided by the span, and the aspect ratio A is defined as the span divided by the mean chord. Hence for a monoplane aerofoil

$$A = \frac{4s^2}{S} \quad \text{and} \quad \frac{S}{4\pi s^2} = \frac{1}{\pi A}.$$

The theoretical formulae, which involve this parameter, will normally be expressed in terms of $\frac{S}{4\pi s^2}$, but the alternative form $\frac{1}{\pi A}$ is useful in a few special cases and for numerical computation.

The induced drag of the aerofoil is

$$\begin{aligned} D_1 &= \int_{-s}^{s} \rho w K \, dy \\ &= \int_{0}^{\pi} 4s^2 \rho V^2 \left(\Sigma n A_n \sin n\theta\right) \left(\Sigma A_n \sin n\theta\right) d\theta \\ &= 2\pi s^2 \rho V^2 \Sigma n A_n^2. \end{aligned}$$

It is convenient to write

$$1 + \delta = \frac{\Sigma n A_n^2}{A_1^2},$$

where δ is a positive quantity whose value is usually small, and then

$$D_1 = \frac{(1 + \delta) L^2}{2\pi s^2 \rho V^2},$$

or

$$C_{D_1} = \frac{S}{4\pi s^2} (1 + \delta) C_L^2.$$

The total drag of the aerofoil is obtained by adding the induced drag and the profile drag. If the aerofoil has a constant aerofoil section across its span and if the effective angle of incidence is also constant, then the profile drag coefficient of each section will have the same value C_{D_0} and the total drag coefficient of the aerofoil will be

$$C_D = C_{D_0} + \frac{S}{4\pi s^2}(1 + \delta)C_L{}^2.$$

More generally the aerofoil section and the effective angle of incidence will vary across the span of the aerofoil and the profile drag coefficient of the aerofoil must be obtained as the value of the integral

$$\frac{1}{S}\int_{-s}^{s} C_{D_0}c\,dy.$$

This refinement, however, is necessary only when the shape of the aerofoil section varies considerably across the span of the aerofoil.

11·22. *Angle of incidence.*

Owing to the normal induced velocity the effective angle of incidence α_0 of any section of an aerofoil is less than the geometrical angle of incidence α, and the aerofoil section gives less lift than it would in two-dimensional motion at the same angle of incidence α. The values of the coefficients A_n, and in particular that of A_1, are determined from the fundamental equation

$$\Sigma A_n \sin n\theta\,(n\mu + \sin\theta) = \mu\alpha\sin\theta$$

as functions of the angle of incidence of the aerofoil, and since A_1 is also proportional to the lift coefficient C_L, a relationship is obtained between the lift coefficient and the angle of incidence of the aerofoil. The slope a of the curve of lift coefficient against angle of incidence determined from this relationship is less than the value a_0 which occurs in the two-dimensional motion of an aerofoil section.

The relationship between the lift coefficient and angle of incidence has a simple form when the aerofoil has a constant angle of incidence across its span. The coefficients A_n are then simply proportional to the angle of incidence α, and the equation

$$A_1 = \frac{S}{4\pi s^2} C_L = \frac{1}{\pi A} C_L$$

gives at once

$$\frac{a}{a_0} = \frac{\pi A}{a_0} \cdot \frac{A_1}{\alpha}.$$

The angle of incidence α of the aerofoil exceeds the angle of incidence α_0 in two-dimensional motion, which would give the same lift coefficient, by the angle

$$\alpha - \alpha_0 = C_L\left(\frac{1}{a} - \frac{1}{a_0}\right),$$

and it is convenient to write this result in a form similar to the equation for the drag coefficient of the aerofoil. Thus

$$\alpha = \alpha_0 + \frac{S}{4\pi s^2}(1 + \tau)C_L,$$

where

$$1 + \tau = \frac{4\pi s^2}{S}\left(\frac{1}{a} - \frac{1}{a_0}\right) = \frac{\alpha}{A_1} - \frac{\pi A}{a_0}.$$

In the more general case of a twisted aerofoil the angle of incidence α varies across the span of the aerofoil and may be expressed in the form

$$\alpha = \bar{a} + f(\theta),$$

where $\bar{a}$ is the angle of incidence at the centre of the aerofoil. The values of the coefficients A_n are then obtained from the fundamental equation in two parts, the first being proportional to and the second independent of $\bar{a}$. The lift coefficient of the aerofoil is therefore of the form

$$C_L = a\bar{a} + k.$$

11·3. *Elliptic loading.*

The lift and induced drag of an aerofoil have been obtained in the forms

$$L = 2\pi s^2 \rho V^2 A_1,$$
$$D_1 = 2\pi s^2 \rho V^2 \Sigma n A_n^2.$$

If an aerofoil of a definite span gives the lift L at the speed V, the coefficient A_1 has a definite value which is independent of the shape of the aerofoil, and the induced drag will be a minimum when all the other coefficients A_n in the series for the circulation are zero. The distribution of circulation across the span of the aerofoil is then simply

$$K = 4sVA_1 \sin \theta = 4sVA_1 \sqrt{1 - \frac{y^2}{s^2}}.$$

The magnitude of the circulation at any point of the span is proportional to the ordinate of an ellipse with the span as major axis, and this type of load distribution is therefore called *elliptic loading*.

The elliptic distribution of circulation or lift across the span of an aerofoil is important, firstly because it leads to the minimum possible induced drag for a given total lift, and secondly because the load grading curves of most aerofoils of conventional shape do not differ greatly from the elliptic form. The results deduced from the hypothesis of elliptic loading are therefore the best which can possibly occur and are also a good first approximation to those actually obtained.

With elliptic loading the normal induced velocity has the constant value

$$w = VA_1 = \frac{S}{4\pi s^2} VC_L$$

across the span of the aerofoil, and the induced drag coefficient of the aerofoil has the value

$$C_{D_1} = \frac{S}{4\pi s^2} C_L^2.$$

If α is the geometrical angle of incidence at any point of the span, the effective angle of incidence will be $\left(\alpha - \dfrac{w}{V}\right)$, and a constant geometrical angle of incidence will imply a constant effective angle of incidence. Hence the lift coefficient also will have the same value for all sections of the aerofoil. But the circulation K round any section is equal to $\frac{1}{2}C_L cV$, and as the circulation varies elliptically across the span, so also will the chord. Thus the elliptic loading will be obtained from a monoplane aerofoil of elliptic plan form and constant angle of incidence. In this case the geometrical and effective angles of incidence are connected by the equation

$$\alpha = \alpha_0 + \frac{S}{4\pi s^2} C_L.$$

Elliptic loading across the span can also be obtained from aerofoils of other plan form by suitable variation of the angle of incidence across the span, but any such twisted aerofoil will give the elliptic loading for one attitude only, since the necessary angle of twist depends on the mean angle of incidence of the aerofoil.

11·31. *Effect of aspect ratio.*

The formulae which have been developed for the angle of incidence and drag coefficient of an elliptic aerofoil can be used to calculate the effect of a change of aspect ratio. If the aspect ratio is reduced from A to A', the changes in the angle of incidence and drag coefficient at a given value of the lift coefficient are respectively

$$\alpha' - \alpha = \frac{1}{\pi}\left(\frac{1}{A'} - \frac{1}{A}\right) C_L,$$

$$C_D' - C_D = \frac{1}{\pi}\left(\frac{1}{A'} - \frac{1}{A}\right) C_L^2.$$

For the standard aspect ratio ($A = 6$) of model experiments the factor $\frac{1}{\pi A}$ has the value 0·053, and if the angle of incidence is measured in degrees this factor becomes 3·05°.

The transformation formula for the drag coefficient applies only to aerofoils with elliptic loading, but the lift distribution curves for rectangular aerofoils and for the majority of aeroplane wings do not differ greatly from the elliptic form, and the transformation formula may therefore be used more generally to calculate the effect of a small change of aspect ratio. The accurate formulae for rectangular and tapered aerofoils are developed in 11·4 and 11·5 respectively.

The transformation formula for the angle of incidence applies only to aerofoils of elliptic plan form with constant angle of incidence across the span and cannot be used for

aerofoils of other plan form except as a rough approximation.
The formula for the angle of incidence also leads to a simple
determination of the slope of the lift curve. If a_0 is the slope
in two-dimensional motion and a is the slope for an elliptic
aerofoil of aspect ratio A, then the formula

$$\alpha = \alpha_0 + \frac{1}{\pi A}\, C_L$$

gives by differentiation

$$\frac{1}{a} = \frac{1}{a_0} + \frac{1}{\pi A},$$

or

$$\frac{a}{\pi} = \frac{A}{1 + \dfrac{\pi}{a_0} A}.$$

Now the theory of an aerofoil in two dimensions gives the
value $a_0 = 2\pi$ approximately and the corresponding values of
a are given in the following table. The values of a, when α is
measured in degrees, are added for comparison with experimental results which are usually quoted in this form.

Table 10.

Slope of lift curve for elliptic aerofoils.

A	∞	10	8	6	4
a (per radian)	6·28	5·24	5·02	4·70	4·18
a (per degree)	0·110	0·092	0·088	0·082	0·074

11·4. *Rectangular aerofoils.*

When the circulation round an aerofoil is expressed in the
form

$$K = 4sV\Sigma A_n \sin n\theta$$

the coefficients A_n must be chosen to satisfy the fundamental
equation

$$\Sigma A_n \sin n\theta \, (n\mu + \sin \theta) = \mu a \sin \theta,$$

where

$$\mu = \frac{a_0 c}{8s}.$$

The successive coefficients $A_1, A_2, A_5 \ldots$ decrease rapidly in
magnitude and it is sufficient to retain only the first three or

four coefficients in order to obtain a good determination of the lift and drag of the aerofoil. The method of solution, when p coefficients are retained, is to determine the coefficients to satisfy the fundamental equation at the p points

$$\theta = \frac{m\pi}{2p}, \qquad m = 1, 2, 3 \dots p.$$

The numerical values given in the subsequent tables are found by retaining the first four coefficients (A_1, A_3, A_5, A_7) and by satisfying the fundamental equation at the four points

$$\theta = 22\tfrac{1}{2}, \qquad 45, \qquad 67\tfrac{1}{2}, \qquad 90 \text{ degrees,}$$

$$\frac{y}{s} = 0\cdot924, \quad 0\cdot707, \quad 0\cdot383, \quad 0.$$

When these four values of θ are inserted in turn in the fundamental equation, four linear equations are obtained to determine the four coefficients. The correct values of μ and α depending on the value of θ must be inserted in these four equations.

The simplest case to consider is that of a rectangular aerofoil* with constant chord c and angle of incidence α across the span. The parameter μ has the constant value

$$\mu = \frac{a_0}{4A},$$

and the coefficients A_1, A_3 ... can be determined conveniently as multiples of $\mu\alpha$. The slope of the curve of lift coefficient against angle of incidence for the aerofoil is then

$$\frac{a}{a_0} = \frac{\pi}{4} \frac{A_1}{\mu\alpha},$$

and the angle of incidence and drag coefficient corresponding to the lift coefficient C_L are

$$\alpha = \alpha_0 + \frac{1}{\pi A}(1 + \tau)\, C_L,$$

$$C_D = C_{D_0} + \frac{1}{\pi A}(1 + \delta)\, C_L{}^2,$$

---

* The solution for a rectangular aerofoil was obtained first by A. Betz by a different and far more laborious process: *Beiträge zur Tragflügeltheorie mit besonderer Berücksichtigung des einfachen rechteckigen Flügels*, Göttingen dissertation, 1919.

where
$$1 + \tau = \frac{1}{\mu}\left(\frac{\mu a}{A_1} - \frac{\pi}{4}\right),$$
$$1 + \delta = \frac{\Sigma n A_n^2}{A_1^2}.$$

The numerical values obtained by this method are given in table 11 and the values of the monoplane coefficients τ and δ are also shown in fig. 85. The values of δ are small

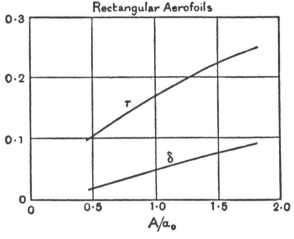

Fig. 85.

and a good first approximation to the induced drag coefficient can be obtained by ignoring δ and by using the elliptic

Table 11.

Rectangular aerofoils.

A/a_0	$A_1/\mu a$	$A_3/\mu a$	$A_5/\mu a$	$A_7/\mu a$	a/a_0	A/a	τ	δ
0·25	·543	·025	·003	·0004	·426	0·58	·05	·007
0·5	·748	·060	·009	·0014	·587	0·85	·10	·019
0·75	·859	·090	·016	·0027	·675	1·11	·14	·034
1·0	·928	·115	·023	·0041	·729	1·37	·17	·049
1·25	·976	·136	·030	·0055	·767	1·63	·20	·063
1·5	1·011	·154	·036	·0070	·794	1·89	·22	·076
1·75	1·038	·169	·042	·0084	·815	2·15	·24	·088

loading formula. The values of τ are not so small and it is necessary always to retain this coefficient in determining the angle of incidence of a rectangular aerofoil.

11·41. *Accuracy of results.*

In order to illustrate the accuracy of the results obtained when only four coefficients (A_1, A_3, A_5, A_7) are retained in the series for the circulation, the calculations have been made in one case $(A = a_0)$ retaining in turn one, two, three and four terms of the series. The results are given in table 12 and show that the values of τ and δ have almost reached their limiting values when four terms are retained. For rapid calculations three terms should be sufficient to give a fair approximation to the true values.

Table 12.

Successive approximations.

No. of terms	$A_1/\mu a$	$A_3/\mu a$	$A_5/\mu a$	$A_7/\mu a$	τ	δ
1	·800	—	—	—	·86	0
2	·917	·084	—	—	·22	·025
3	·926	·110	·016	—	·18	·044
4	·928	·115	·023	·004	·17	·049

11·42. *Effect of aspect ratio.*

The conversion formulae for a change of aspect ratio from A to A' for rectangular aerofoils are

$$\alpha' - \alpha = \frac{1}{\pi}\left(\frac{1+\tau'}{A'} - \frac{1+\tau}{A}\right)C_L,$$

$$C_D' - C_D = \frac{1}{\pi}\left(\frac{1+\delta'}{A'} - \frac{1+\delta}{A}\right)C_L^2.$$

Table 11 or fig. 85 gives the values of τ and δ in terms of A/a_0. The relationship between A/a and A/a_0 can be obtained from the same table or from fig. 86, and approximates closely to the linear law

$$\frac{A}{a} = 0·33 + 1·04\frac{A}{a_0}.$$

The experimental determination of the characteristics of a rectangular aerofoil of aspect ratio A gives the value of A/a and

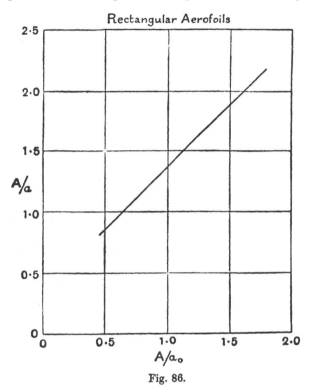

Fig. 86.

the corresponding values of A/a_0, and τ and δ can be obtained from the figures. The value of a_0 is then known and the values of τ' and δ' for any other aspect ratio A' can be found at once from fig. 85.

If $a_0 = 2\pi$, the values of τ and δ for the standard aspect ratio 6 are

$$\tau = 0.163, \qquad \delta = 0.046,$$

and then, if the angle of incidence is expressed in degrees,

$$\alpha = \alpha_0 + 3.55^\circ C_L,$$
$$C_D = C_{D_0} + 0.0555 C_L^2.$$

These numerical values should be compared with the corresponding values $3 \cdot 05°$ and $0 \cdot 053$ for an elliptic aerofoil of the same aspect ratio.

Finally, the slope of the lift curve for rectangular aerofoils of different aspect ratio, on the assumption that $a_0 = 2\pi$, is given in table 13. These values are all slightly less than the corresponding values for elliptic aerofoils given in table 10.

Table 13.

Slope of lift curve for rectangular aerofoils.

A	∞	10	8	6	4	2
a (per radian)	6·28	5·04	4·84	4·54	4·04	3·04
a (per degree)	0·110	0·088	0·084	0·080	0·070	0·054

11·43. *Pitching moment.*

The relationship between the moment coefficient and the lift coefficient of a uniform rectangular aerofoil is the same as for the aerofoil section in two-dimensional motion and is of the form

$$C_M = m_0 + m_1 C_L,$$

where m_0 and m_1 are both negative in general. For, if this equation represents the moment coefficient of the aerofoil section in two-dimensional motion, the pitching moment of the rectangular aerofoil about its leading edge will be

$$C_M \cdot \tfrac{1}{2}\rho V^2 S c = \frac{1}{2}\int_{-s}^{s}\left(m_0 + m_1 \frac{2K}{cV}\right)c^2\rho V^2 dy,$$

where $2K/cV$ has been written for the lift coefficient of the aerofoil section. Putting

$$y = -s\cos\theta,$$
$$S = 2sc,$$
$$K = 4sV\Sigma A_n \sin n\theta,$$

the equation gives

$$C_M = \frac{1}{2}\int_0^{\pi}\left\{m_0 + \frac{8sm_1}{c}\Sigma A_n \sin n\theta\right\}\sin\theta\, d\theta$$
$$= m_0 + m_1 \frac{2\pi s}{c}A_1$$
$$= m_0 + m_1 C_L,$$

since $$A_1 = \frac{S}{4\pi s^2} C_L = \frac{c}{2\pi s} C_L.$$

11·5. *Tapered aerofoils.*

Another important type of aerofoil is that in which the chord decreases uniformly from a maximum value c_0 at the

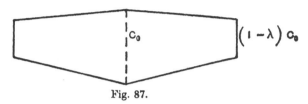

Fig. 87.

centre of the aerofoil to $c_0 (1 - \lambda)$ at the tips. In general this change of chord is associated with a change of the aerofoil section, but it will be assumed in the first place that the aerofoil has no aerodynamic twist, i.e. that the angle of incidence measured from the no lift line of the sections is constant across the span of the aerofoil.

Solutions are obtained as before from the fundamental equation

$$\Sigma A_n \sin n\theta \, (n\mu + \sin \theta) = \mu a \sin \theta,$$

and the solution follows the same lines except that μ now has different values at the four typical points

$$\theta = 22\tfrac{1}{2}, \quad 45, \quad 67\tfrac{1}{2}, \quad 90 \text{ degrees.}$$

For this range $c = c_0 (1 - \lambda \cos \theta),$

$$\mu = \mu_0 (1 - \lambda \cos \theta),$$

where $\mu_0 = \dfrac{a_0 c_0}{8s}.$

The area of the aerofoil is

$$S = (2 - \lambda) s c_0$$

and the aspect ratio is

$$A = \frac{4s}{(2 - \lambda) c_0} = \frac{a_0}{2 (2 - \lambda) \mu_0}.$$

Numerical results illustrating the effect of taper are given in table 14 for the aspect ratio $A = a_0$, and the corresponding values of the monoplane coefficients τ and δ are shown in fig. 88. It appears that the best results are obtained when

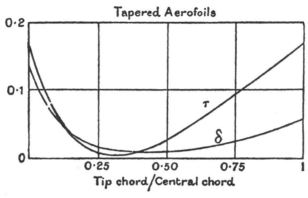

Fig. 88.

the tip chord is from one-third to one-half of the central chord, for it is desirable that τ and δ should both be as small as possible.

Table 14.

Tapered aerofoils $(A = a_0)$.

λ	A_1/a	A_3/a	A_5/a	A_7/a	a/a_0	τ	δ
0	·232	·029	·006	·001	729	·17	·049
0·25	·236	·020	·008	0	742	10	·026
0·50	·240	·007	·010	−·001	·754	·03	·011
0·75	·241	−·012	·010	−·002	·757	·01	·016
1·00	·232	−·050	·002	−·004	729	·17	·141

11·6. *Twisted aerofoils.*

If the zero lift lines of the sections of an aerofoil are not parallel to one another along the span, the aerofoil is twisted aerodynamically. This twist may be due to variation of camber along the span, with the chord lines of the sections remaining parallel to one another, or it may be due to variation of the setting of the chord lines of the sections. If the incidence at the tips is less than the incidence at the centre the aerofoil is said to have a "wash-out" towards the tips.

As an illustration of the method of calculating the characteristics of a twisted aerofoil, consider a rectangular aerofoil of constant section whose geometrical angle of incidence decreases uniformly from the centre to the tips. Let $\bar{a}$ be the angle of incidence at the centre of the aerofoil and ϵ the decrease from the centre to either tip. Then for the port half of the aerofoil the angle of incidence is

$$a = \bar{a} - \epsilon \cos \theta,$$

and the fundamental equation for the coefficients A_1, A_3 ... of the series for the circulation becomes

$$\Sigma A_n \sin n\theta \, (n\mu + \sin \theta) = \mu \sin \theta \, (\bar{a} - \epsilon \cos \theta).$$

The solution for the first four coefficients of the series proceeds as before, but each coefficient is now determined in two parts, the first being proportional to $\mu\bar{a}$ and the second to $\mu\epsilon$. Numerical values for the case $A = a_0$ are

$$A_1 = 0\cdot928\mu\bar{a} - 0\cdot408\mu\epsilon,$$
$$A_3 = 0\cdot115\mu\bar{a} - 0\cdot242\mu\epsilon,$$
$$A_5 = 0\cdot023\mu\bar{a} + 0\cdot010\mu\epsilon,$$
$$A_7 = 0\cdot004\mu\bar{a} - 0\cdot023\mu\epsilon.$$

The lift coefficient of the twisted aerofoil is

$$C_L = \frac{4\pi s^2}{S} A_1 = 4\cdot56\bar{a} - 2\cdot02\epsilon,$$

whose slope is the same as for the corresponding untwisted rectangular aerofoil.

The induced drag coefficient of the aerofoil may still be written in the form

$$C_{D_1} = \frac{S}{4\pi s^2}(1 + \delta) C_L^2,$$

but the coefficient δ now varies with the angle of incidence of the aerofoil. Thus if ϵ is equal to $0\cdot1$ radian ($5\cdot7°$), the characteristics of the aerofoil at different angles of incidence are as follows:

$\bar{a}$	C_L	δ
0·10	0·256	0·205
0·15	0·484	0·027
0·20	0·712	0·009
0·25	0·940	0·003

These values of δ should be compared with the value 0·049 for the corresponding untwisted rectangular aerofoil, and it appears that the twisted aerofoil with "wash out" towards the tips has the lower induced drag except at very low lift coefficients.

The more general case of a twisted tapered aerofoil can be solved along similar lines, the only modification being that the parameter μ must now be regarded as a function of the coordinate θ as in 11·5.

11·7. *Load grading curves.*

The solution of the problem of a monoplane aerofoil in the form of a Fourier series for the circulation can also be used to determine the shape of the load grading curve across the span of the aerofoil, since the lift on any element of the span of an aerofoil is proportional to the circulation round that element. In general the first four terms only of the Fourier series have been determined and the corresponding load grading curve is of a sinuous nature. The solution, however, is exact only at the four points of the semi-span ($\theta = 22\frac{1}{2}$, 45, 67$\frac{1}{2}$, 90 degrees) which are used in determining the coefficients of the Fourier series, and the load grading curve should therefore be drawn as a smooth curve through the values determined at these points.

Fig. 89 shows the load grading curves for various aerofoil shapes determined in this manner, the scale of the ordinates being chosen so that each aerofoil carries the same total load. A wing of elliptic plan form gives a load grading curve of the

same form, and an aerofoil of any other form gives a load
grading curve whose form is intermediate between that of

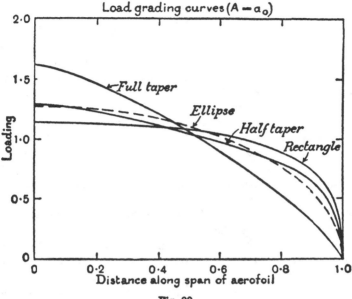

Fig. 89.

the aerofoil and that of the ellipse. Aerodynamically the
merit of an aerofoil is measured by the closeness with which
the load grading curve approximates to the elliptic form.

CHAPTER XII

THE FLOW ROUND AN AEROFOIL

12·1. *The flow pattern.*

The deviation of the velocity at any point of the fluid from the undisturbed velocity V is due to the vortex system created by the aerofoil and can be calculated as the velocity field of this vortex system. The general nature of the vortex system, comprising the circulation round the aerofoil and the trailing vortices which spring from its trailing edge, has been discussed in 10·2, and the analysis of chapter XI provides a method of determining the strength of the vortex system associated with any monoplane aerofoil. The analysis is based on the assumption that the aerofoil can be replaced by a lifting line, and calculations based on this assumption will clearly be inadequate to determine the flow in the immediate neighbourhood of the aerofoil where the shape of the aerofoil sections will modify the form of the flow pattern. Also in the neighbourhood of the vortex wake it is necessary to consider the tendency of the trailing vortex sheet to roll up into a pair of finite vortices. Apart from these two limitations it is possible to obtain a satisfactory account of the flow pattern round an aerofoil from the simple assumption of a lifting line and of straight line vortices extending indefinitely down stream. Finally, at large distances from the aerofoil and its wake, the velocity field will depend only on the lift carried by the aerofoil and will be independent of the span of the aerofoil and of the shape of the load grading curve.

In determining the flow pattern round an aerofoil the standard system of axes will be used with origin at the mid-point of the aerofoil. The axis of x extends forwards in the direction of motion of the aerofoil relative to the air, the axis of y is along the span of the aerofoil to starboard, and the axis of z is normal to the first two axes as in fig. 90. The velocity field of the vortex system, which represents the disturbance created by the aerofoil in the uniform stream V, will be expressed by the velocity components (u, v, w) parallel to these axes.

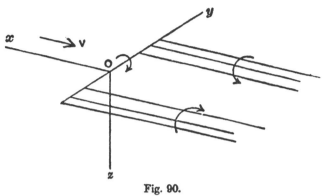

Fig. 90.

It follows at once from the simple form assumed for the vortex system that the longitudinal velocity component u depends only on the circulation round the aerofoil, and that the lateral velocity component v depends only on the system of trailing vortices, whereas the normal velocity component w depends on the complete vortex system.

The flow pattern will be examined first on the assumption of uniform loading across the span of the aerofoil and attention will be directed mainly to the normal velocity component w. The effect of other forms of load grading curve will be considered only in a few regions of special importance.

12·2. *Uniform loading.*

The vortex system of an aerofoil with uniform loading across the span consists of the aerofoil AA' and the two straight trailing vortices AB and $A'B'$, the strength of the circulation being K for the whole system. The typical point P at which the induced velocity is to be determined will be chosen for convenience as in fig. 91 with negative values of the coordinates x and z. Let PL, PM and PN be the perpendiculars from P to the plane Oxy and to the lines Oy and AB respectively. Then

$$ML = -x = x',$$
$$OM = y,$$
$$PL = -z = z',$$
$$OA = OA' = s.$$

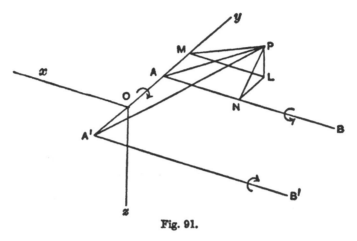

Fig. 91.

The induced velocity at P due to the circulation K round the aerofoil AA' is normal to the plane PMO and has the value

$$q_1 = \frac{K}{4\pi PM}(\cos PA'A + \cos PAA')$$

$$= \frac{K}{4\pi\sqrt{x'^2+z'^2}}\left\{\frac{y+s}{\sqrt{x'^2+z'^2+(y+s)^2}} - \frac{y-s}{\sqrt{x'^2+z'^2+(y-s)^2}}\right\},$$

and the components of this velocity are

$$u_1 = -\frac{K}{4\pi}\frac{z'}{x'^2+z'^2}\left\{\frac{y+s}{\sqrt{x'^2+z'^2+(y+s)^2}} - \frac{y-s}{\sqrt{x'^2+z'^2+(y-s)^2}}\right\},$$

$$v_1 = 0,$$

$$w_1 = \frac{K}{4\pi}\frac{x'}{x'^2+z'^2}\left\{\frac{y+s}{\sqrt{x'^2+z'^2+(y+s)^2}} - \frac{y-s}{\sqrt{x'^2+z'^2+(y-s)^2}}\right\}.$$

The induced velocity at P due to the circulation K round the trailing vortex AB is normal to the plane PAB and has the value

$$q_2 = \frac{K}{4\pi \overline{PN}} (1 + \cos PAB)$$

$$= \frac{K}{4\pi \sqrt{z'^2 + (y - s)^2}} \left\{ 1 + \frac{x'}{\sqrt{x'^2 + z'^2 + (y - s)^2}} \right\},$$

and the components of this velocity are

$$u_2 = 0,$$

$$v_2 = - \frac{K}{4\pi} \frac{z'}{z'^2 + (y - s)^2} \left\{ 1 + \frac{x'}{\sqrt{x'^2 + z'^2 + (y - s)^2}} \right\},$$

$$w_2 = - \frac{K}{4\pi} \frac{y - s}{z'^2 + (y - s)^2} \left\{ 1 + \frac{x'}{\sqrt{x'^2 + z'^2 + (y - s)^2}} \right\}.$$

Similarly the velocity components at P due to the circulation K round the trailing vortex $A'B'$ are

$$u_3 = 0,$$

$$v_3 = \frac{K}{4\pi} \frac{z'}{z'^2 + (y + s)^2} \left\{ 1 + \frac{x'}{\sqrt{x'^2 + z'^2 + (y + s)^2}} \right\},$$

$$w_3 = \frac{K}{4\pi} \frac{y + s}{z'^2 + (y + s)^2} \left\{ 1 + \frac{x'}{\sqrt{x'^2 + z'^2 + (y + s)^2}} \right\}.$$

The components of the induced velocity at P, due to the aerofoil and its trailing vortices, are obtained as the sums of the expressions given above:

$$u = u_1,$$
$$v = v_2 + v_3,$$
$$w = w_1 + w_2 + w_3.$$

The detailed examination of these expressions will be confined to the regions of special interest, i.e. to points lying in the lateral plane ($x = 0$) and to points on the axis of x.

The induced velocity is proportional to the circulation K, which is related to the lift of the aerofoil by the equation

$$2s\rho V K = L = C_L \cdot \tfrac{1}{2}\rho V^2 S,$$

or $$\frac{K}{\pi s} = \frac{S}{4\pi s^2} V C_L.$$

This last expression is equal to the normal induced velocity w_0 at an aerofoil with elliptic loading (cf. 11·3) and it is convenient to express the induced velocity components at the general point P as multiples of this velocity w_0.

12·21. *The lateral plane.*

The longitudinal component of the induced velocity at a point of the lateral plane ($x = 0$) is

$$u = \frac{K}{4\pi z} \left\{ \frac{y+s}{\sqrt{z^2+(y+s)^2}} - \frac{y-s}{\sqrt{z^2+(y-s)^2}} \right\},$$

and for points on the axis of z the expression reduces further to

$$u = \frac{K}{2\pi} \frac{s}{z\sqrt{z^2+s^2}}$$

$$= \frac{1}{2z}\frac{s^2}{\sqrt{z^2+s^2}}\frac{SVC_L}{4\pi s^2}.$$

The longitudinal component of the velocity of the air relative to the aerofoil is $(V - u)$ and is therefore increased above the aerofoil, where z and u are negative, and is decreased below it. The variation of the correction to the longitudinal velocity is shown by the following numerical values:

$$\frac{z}{s} = \quad \tfrac{1}{4} \qquad \tfrac{1}{2} \qquad 1 \qquad 2,$$

$$\frac{u}{w_0} = 1\cdot42 \qquad 0\cdot62 \qquad 0\cdot35 \qquad 0\cdot11.$$

For an aerofoil of aspect ratio 6 the velocity w_0 has the value $0\cdot053 VC_L$. Thus the correction to the longitudinal velocity is only of the order of 2 % at a depth below the aerofoil equal to the semi-span when the lift coefficient has the large value 0·5.

12·22. The lateral and normal components of the induced velocity at a point of the lateral plane ($x = 0$) are respectively

$$v = \frac{K}{2\pi} \frac{2syz}{(y^2+z^2+s^2)^2 - 4y^2s^2},$$

$$w = -\frac{K}{2\pi} \frac{s(y^2-z^2-s^2)}{(y^2+z^2+s^2)^2 - 4y^2s^2}.$$

The denominator of these two expressions is essentially positive and hence the lateral component of the velocity is directed inwards above the aerofoil and outwards below it, while the sign of the normal component depends on that of $(y^2 - z^2 - s^2)$.

At large distances from the aerofoil the expressions tend to the values

$$v = \frac{K}{2\pi}\frac{2syz}{(y^2+z^2)^2} = \frac{2yz}{(y^2+z^2)^2}\frac{SVC_L}{8\pi},$$

$$w = -\frac{K}{2\pi}\frac{s(y^2-z^2)}{(y^2+z^2)^2} = -\frac{y^2-z^2}{(y^2+z^2)^2}\frac{SVC_L}{8\pi}.$$

Also on the axis of z the normal component of the induced velocity is

$$w = \frac{K}{2\pi}\frac{s}{z^2+s^2} = \frac{1}{2}\frac{s^2}{z^2+s^2}\frac{SVC_L}{4\pi s^2},$$

and has the following numerical values:

$$\frac{z}{s} = \tfrac{1}{4} \quad \tfrac{1}{2} \quad 1 \quad 2,$$

$$\frac{w}{w_0} = 0.45 \quad 0.35 \quad 0.25 \quad 0.10.$$

For an aerofoil of aspect ratio 6 the velocity w_0 has the value $0.053VC_L$, and at a depth below the aerofoil equal to the semi-span the downward velocity is $0.0132VC_L$, representing an angle of downwash of $0.75°C_L$.

12.23. *The longitudinal axis.*

At a point on the axis of x ($y = z = 0$), the longitudinal and lateral components of the induced velocity are zero and the normal component has the value

$$w = -\frac{K}{2\pi}\frac{s}{x\sqrt{x^2+s^2}} + \frac{K}{2\pi s}\left\{1 - \frac{x}{\sqrt{x^2+s^2}}\right\}$$

$$= \frac{K}{2\pi s}\left\{1 - \frac{\sqrt{x^2+s^2}}{x}\right\}.$$

In front of the aerofoil the normal velocity is negative and the air is flowing upwards to meet the aerofoil. Behind the aerofoil the normal velocity is positive, and at a distance l behind the aerofoil the angle of downwash ϵ is

$$\epsilon = \frac{w}{V} = \frac{K}{2\pi s V}\left\{1 + \frac{\sqrt{l^2 + s^2}}{l}\right\}$$

$$= \frac{1}{2}\left\{1 + \frac{\sqrt{l^2 + s^2}}{l}\right\}\frac{S}{4\pi s^2}C_L.$$

Writing
$$\epsilon_0 = \frac{w_0}{V} = \frac{S}{4\pi s^2}C_L$$

for the angle of downwash corresponding to the standard induced velocity w_0, the expression for the angle of downwash behind the aerofoil becomes

$$\epsilon = \frac{1}{2}\left\{1 + \frac{\sqrt{l^2 + s^2}}{l}\right\}\epsilon_0,$$

and has the following numerical values:

$\dfrac{l}{s} =$	$\tfrac{1}{4}$	$\tfrac{1}{2}$	1	2,
$\dfrac{\epsilon}{\epsilon_0} =$	2·08	1·40	1·21	1·06.

12·3. *Elliptic loading.*

The flow pattern has been considered hitherto on the assumption that the circulation has a constant value across the span of the aerofoil. This condition is not satisfied by any aerofoil, but the actual distribution of circulation across the span can always be built up by superimposing a number of simple "horseshoe" systems (cf. 10·21 and fig. 77). In order to derive the corresponding values of the induced velocity at any point it is necessary to replace the length s in the formulae of 12·2 by a coordinate η measured along the span of the aerofoil, to replace the circulation K by $-\dfrac{dK}{d\eta}\,d\eta$, and to integrate the expressions from $\eta = 0$ to $\eta = s$. This integration is however extremely complex in general.

In the case of elliptic loading (cf. 11·3), the circulation at any point of the aerofoil is

$$K = 4sVA_1\sqrt{1 - \frac{\eta^2}{s^2}},$$

and the normal induced velocity at the aerofoil is

$$w_0 = VA_1.$$

Hence
$$K = 4w_0\sqrt{s^2 - \eta^2}$$

$$-\frac{dK}{d\eta} = \frac{4w_0\eta}{\sqrt{s^2 - \eta^2}}.$$

12·31. *The normal axis.*

The normal induced velocity at a point on the axis of z in the case of uniform loading has been obtained in 12·22 in the form

$$w = \frac{K}{2\pi}\frac{s}{z^2 + s^2},$$

and hence for elliptic loading

$$w = \int_0^s \frac{1}{2\pi}\frac{4w_0\eta}{\sqrt{s^2 - \eta^2}}\frac{\eta}{z^2 + \eta^2}\,d\eta,$$

which can be integrated simply by means of the substitution $\eta = s\sin\theta$ and gives

$$w = w_0\left(1 - \frac{z}{\sqrt{z^2 + s^2}}\right).$$

The numerical values given by this formula are:

$\frac{z}{s} =$	$\frac{1}{4}$	$\frac{1}{2}$	1	2
$\frac{w}{w_0} =$	0·68	0·45	0·29	0·11

and a comparison with the values given in 12·22 for uniform loading shows that at a depth below the aerofoil equal to the semi-span the difference in the induced velocity is $0·04w_0$. This difference is negligible, for it corresponds to a difference of only $0·12°C_L$ in the angle of downwash for an aerofoil of aspect ratio 6.

When the distance from the aerofoil is large, the normal induced velocity tends to the value

$$w = \frac{1}{2}\frac{s^2}{z^2}w_0 = \frac{S}{8\pi z^2}VC_L,$$

both for uniform and for elliptic loading, illustrating the general theorem that at large distances from an aerofoil and its wake the induced velocity depends only on the total lift and is independent both of the span and of the form of the load grading curve.

12·32. *The lateral plane.*

In the case of elliptic loading the flow in the lateral plane ($x = 0$) can be obtained from the fact that the normal induced velocity has a constant value w_0 across the span of the aerofoil. The flow is therefore identical with the two dimensional motion caused by a line of length $2s$ moving normally to itself with the velocity w_0. The stream function of this flow is given in 6·23 and the flow is illustrated in fig. 44. Writing

$$y = s \sin \lambda \cosh \mu,$$
$$z = s \cos \lambda \sinh \mu,$$

the normal velocity w can be shown to be

$$w = w_0 \left\{ 1 - \frac{\sinh \mu \cosh \mu}{\cosh^2 \mu - \sin^2 \lambda} \right\}.$$

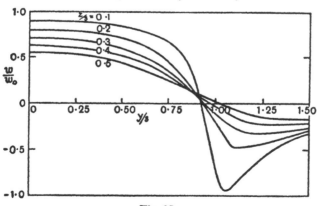

Fig. 92.

The numerical values of w deduced from this formula for points in the neighbourhood of the aerofoil are given in table 15 below and are shown in fig. 92.

Far from the aerofoil the following limiting values are obtained:

$$y = \tfrac{1}{2} s e^{\mu} \sin \lambda,$$
$$z = \tfrac{1}{2} s e^{\mu} \cos \lambda,$$
$$\frac{w}{w_0} = 2 e^{-2\mu} (\cos^2 \lambda - \sin^2 \lambda)$$
$$= -\frac{s^2}{2} \frac{y^2 - z^2}{(y^2 + z^2)^2},$$

or
$$w = -\frac{y^2 - z^2}{(y^2 + z^2)^2} \frac{S V C_L}{8\pi},$$

which is the same as the limiting value obtained in 12·22 from the assumption of uniform loading.

Table 15.
Values of w/w_0.

$y/s =$	0	0·25	0·50	0·75	0·90	1·10	1·25	1·50
$z/s=$ 0	1·00	1·00	1·00	1·00	1·00	−1·40	−0·67	−0·34
0·1	0·90	0·89	0·85	0·68	0·20	−0·92	−0·60	−0·33
0·2	0·80	0·79	0·71	0·47	0·07	−0·49	−0·45	−0·30
0·3	0·71	0·69	0·60	0·35	0·07	−0·27	−0·32	−0·25
0·4	0·63	0·60	0·50	0·29	0·09	−0·14	−0·21	−0·21
0·5	0·55	0·53	0·43	0·25	0·10	−0·07	−0·14	−0·16

12·33. The longitudinal axis.

The angle of downwash at a point on the axis of x in the case of uniform loading has been obtained in 12·23 in the form

$$\epsilon = \frac{K}{2\pi s V} \left\{ 1 + \frac{\sqrt{l^2 + s^2}}{l} \right\},$$

and hence for elliptic loading

$$\epsilon = \int_0^s \frac{1}{2\pi V} \frac{4 w_0}{\sqrt{s^2 - \eta^2}} \left\{ 1 + \frac{\sqrt{l^2 + \eta^2}}{l} \right\} d\eta$$
$$= \frac{2}{\pi} \epsilon_0 \int_0^s \left\{ 1 + \frac{\sqrt{l^2 + \eta^2}}{l} \right\} \frac{d\eta}{\sqrt{s^2 - \eta^2}}.$$

Substituting $\eta = s \cos \theta,$

$$l^2 = \frac{1 - k^2}{k^2} s^2,$$

this integral becomes

$$\frac{\epsilon}{\epsilon_0} = \frac{2}{\pi} \int_0^{\frac{\pi}{2}} \left\{ 1 + \sqrt{\frac{1 - k^2 \sin^2 \theta}{1 - k^2}} \right\} d\theta$$

$$= 1 + \frac{2}{\pi} \frac{E}{\sqrt{1 - k^2}},$$

where E is the complete elliptic integral

$$E = \int_0^{\frac{\pi}{2}} \sqrt{1 - k^2 \sin^2 \theta} \, d\theta.$$

Numerical values of the angle of downwash deduced from this formula are:

$\frac{l}{s} =$	$\frac{1}{3}$	$\frac{2}{3}$	1	2
$\frac{\epsilon}{\epsilon_0} =$	3·23	2·43	2·22	2·06

These values are consistently larger than those deduced from the assumption of uniform loading and tabulated in 12·23. Moreover, they tend to the limit $2\epsilon_0$ instead of ϵ_0 as l tends to infinity. Neither of these sets of values can be regarded as satisfactory. They are based on the assumption that the trailing vortices extend backwards indefinitely as straight lines, and to obtain a reliable estimate of the angle of downwash behind an aerofoil it is necessary to take account of the fact that the sheet of trailing vortices is unstable and rolls up into a pair of vortices.

12·4. *Angle of downwash.*

The formulae for the normal induced velocity w on the longitudinal axis, which have been developed in the preceding paragraphs 12·23 and 12·33, are based on the assumption that the trailing vortices extend backwards indefinitely as straight lines. Actually the trailing vortex sheet is unstable and rolls up into a pair of vortices whose distance

apart $(2s')$ is rather less than the span $(2s)$ of the aerofoil.
The strength of each of the resulting vortices will clearly be
equal to the magnitude of
the circulation $\bar{K}$ round
the centre of the aerofoil.
At points distant from the
aerofoil and its wake this
modified vortex system will
be equivalent to that of an
aerofoil of span $2s'$ with
uniform circulation $\bar{K}$ and

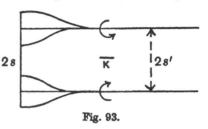

Fig. 93.

hence the distance s' can be determined from the equation

$$L = 2s'\rho V \bar{K}.$$

Now any form of load distribution across the span of an
aerofoil can be represented as in 11·2 by the series

$$K = 4sV \Sigma A_n \sin n\theta,$$

and then
$$L = 2\pi s^2 \rho V^2 A_1,$$

$$\bar{K} = 4sV (A_1 - A_3 + A_5 - \ldots).$$

Hence
$$\frac{s'}{s} = \frac{L}{2s\rho V \bar{K}} = \frac{\pi}{4} \frac{A_1}{(A_1 - A_3 + A_5 - \ldots)}.$$

The normal induced velocity w at a point on the longitudinal axis at some distance behind the aerofoil will be
estimated more accurately by the use of this modified vortex
system than by the assumption which ignores the rolling up
of the sheet of trailing vortices. Thus the vortex system is
assumed to be that of an aerofoil of span $2s'$ with constant
circulation $\bar{K}$, and the normal induced velocity, according to
12·23, becomes

$$w = \frac{\bar{K}}{2\pi s'}\left(1 + \frac{\sqrt{l^2 + s'^2}}{l}\right)$$

$$= \frac{L}{4\pi s'^2 \rho V}\left(1 + \frac{\sqrt{l^2 + s'^2}}{l}\right).$$

The angle of downwash is therefore

$$\epsilon = \frac{1}{2}\frac{s^2}{s'^2}\left(1 + \frac{\sqrt{l^2 + s'^2}}{l}\right)\epsilon_0,$$

where
$$\epsilon_0 = \frac{S}{4\pi s^2} C_L,$$

and the limiting value of the angle of downwash as the distance l tends to infinity is
$$\epsilon = \frac{s^2}{s'^2} \epsilon_0.$$

The values of $\frac{s'}{s}$ for rectangular aerofoils can be calculated from the results given in table 11 of chapter XI and are recorded in table 16 below together with the corresponding values of $\frac{\epsilon}{\epsilon_0}$ at the point $l = s$ and the limiting values as l tends to infinity.

Finally, the rate of change of the angle of downwash with angle of incidence is
$$\frac{d\epsilon}{d\alpha} = \frac{1}{2} \frac{s^2}{s'^2} \left(1 + \frac{\sqrt{l^2 + s'^2}}{l}\right) \frac{S}{4\pi s^2} \frac{dC_L}{d\alpha}$$
$$= \frac{\epsilon}{\epsilon_0} \frac{a}{\pi A},$$

and the value of this expression at the point $l = s$ is also given in table 16.

The definition of the length l is somewhat uncertain since the calculations are based on the assumption that the aerofoil can be replaced by a lifting line and the position of this line

Table 16.

Angle of downwash.

Aerofoil	s'/s	ϵ/ϵ_0(limit)	ϵ/ϵ_0 ($l=s$)	$d\epsilon/d\alpha$ ($l=s$)
Ellipse	0·785	1·62	1·84	0·44 ($A = a_0$)
Rectangle: $A/a_0 = 0.5$	0·844	1·40	1·62	0·61
0·75	0·862	1·35	1·57	0·45
1·0	0·875	1·31	1·52	0·35
1·25	0·887	1·27	1·48	0·29
1·5	0·896	1·25	1·46	0·25
1·75	0·903	1·23	1·44	0·21

in the aerofoil has not been defined. Clearly, however, the
lifting line should pass through the centre of pressure of the
aerofoil and the length l is therefore the distance behind the
centre of pressure of the aerofoil.

The variation of the angle of downwash with the distance
behind the aerofoil is shown in fig. 94, where the curves are
drawn for an elliptic aerofoil and for a rectangular aerofoil
of aspect ratio $A = a_0$. The broken curves represent the
corresponding values for uniform and elliptic loadings when
the rolling up of the sheet of trailing vortices is ignored.

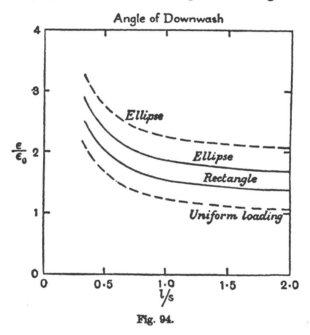

Fig. 94.

The results given in this chapter refer in all cases to a
monoplane aerofoil, but the flow pattern for a biplane system
can be derived by adding the effects due to the two separate
aerofoils. In particular the angle of downwash behind a
biplane system consisting of two rectangular aerofoils of
aspect ratio A will be nearly double that behind a monoplane

aerofoil of the same aspect ratio at the same lift coefficient. On the other hand, the value of $\frac{d\epsilon}{d\alpha}$ will not be doubled owing to the decrease in the value of $\frac{dC_L}{d\alpha}$ for a biplane. Using the numerical values given in 11·42 and 13·24 for rectangular aerofoils of aspect ratio 6, the value of $\frac{d\epsilon}{d\alpha}$ is found to be roughly 0·35 for the monoplane and 0·55 for the biplane.

CHAPTER XIII

BIPLANE AEROFOILS

13·1. *Two-dimensional motion.*

The problem of the two dimensional flow past a pair of aerofoils forming a biplane system is very complex and complete solutions have been obtained only for the case when the aerofoil sections are straight lines. A brief outline only is given here to indicate the method of analysis and the general nature of the results obtained.

13·11. *Tandem aerofoils.*

Consider first a tandem system formed by two equal segments AB and $A'B'$ of the real axis with their extremities at the points $x = \pm p, \pm q$. The most general type of irrotational flow past this system can be expressed in the form

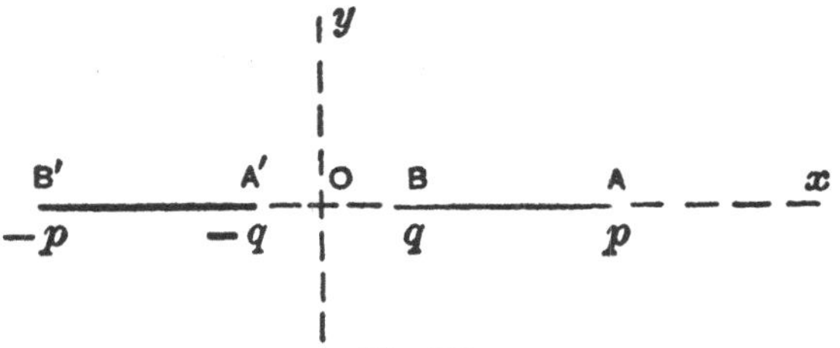

Fig. 95.

$$u - iv = \frac{dw}{dz} = U + U' \frac{z^2 - m^2}{\sqrt{(p^2 - z^2)(z^2 - q^2)}}$$
$$+ \frac{C}{\pi} \frac{z}{\sqrt{(p^2 - z^2)(z^2 - q^2)}} + \frac{C'}{\pi} \frac{n}{\sqrt{(p^2 - z^2)(z^2 - q^2)}},$$

where the four terms represent respectively the flow due to a uniform velocity U parallel to the axis of x, a uniform velocity U' parallel to the axis of y, equal circulation C round each aerofoil, and positive circulation C' round the first aerofoil and negative circulation C' round the second aerofoil. The quantities m and n are constants whose values are determined later as functions of p and q.

The general expression represents a possible irrotational

motion since the potential function w is a function of the complex variable z, it has the correct limiting value as z tends to infinity, and it gives zero normal velocity at the surface of the aerofoils and finite velocity at all points except the ends of the aerofoils. The sign of the radical

$$\sqrt{(p^2 - z^2)(z^2 - q^2)}$$

must be chosen in accordance with the vectorial interpretation of the expression. The sign is positive on the lower surface of AB and on the upper surface of $A'B'$, and is negative on the opposite surfaces.

The value of the constant m is determined from the condition that there is no circulation round either aerofoil for the simple vertical flow U', and the value of the constant n from the condition that the circulation round the aerofoil AB is equal to C' for the fourth type of flow. These values are

$$m = p\sqrt{\frac{E}{K}},$$

$$n = \frac{\pi}{2}\frac{p}{K},$$

where E and K are the complete elliptic integrals for the modulus k, defined by the equations

$$k^2 = \frac{p^2 - q^2}{p^2},$$

$$E = \int_0^1 \sqrt{\frac{1 - k^2 x^2}{1 - x^2}}\, dx,$$

$$K = \int_0^1 \frac{dx}{\sqrt{(1 - x^2)(1 - k^2 x^2)}}.$$

The points $x = \pm m$ are the stagnation points of the simple vertical flow on the surface of the aerofoils.

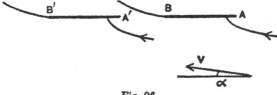

Fig. 96.

To obtain the flow for angle of incidence α which leaves the trailing edges B and B' smoothly, it is necessary to write
$$U = -V\cos\alpha, \qquad U' = V\sin\alpha,$$
and to choose the values of C and C' so that the numerator over the radical in the general expression for $(u - iv)$ is zero at the points B and B'. Hence
$$C = \pi(p - q)V\sin\alpha,$$
$$C' = 2(pE - qK)V\sin\alpha.$$

The resultant force on the tandem system is the lift
$$L = 2\rho VC = 2\pi(p - q)\rho V^2 \sin\alpha,$$
corresponding to the total circulation $2C$, and the tandem system therefore gives the same total lift as a single aerofoil of the same total chord $2(p - q)$.

The forces on the individual aerofoils can be determined by evaluating the integral
$$X - iY = \frac{i\rho}{2}\int\left(\frac{dw}{dz}\right)^2 dz$$
round the surface of each aerofoil in turn. By applying this method it is found that the front aerofoil AB experiences a greater lift force than the rear aerofoil $A'B'$, and that the rear aerofoil experiences a drag force which is balanced by an equal forward force on the front aerofoil.

13·12. *The unstaggered biplane.*

An unstaggered biplane system, formed by two equal parallel lines, can be derived from the tandem system by the conformal transformation
$$\frac{d\zeta}{dz} = \frac{z^2 - m^2}{\sqrt{(p^2 - z^2)(z^2 - q^2)}},$$

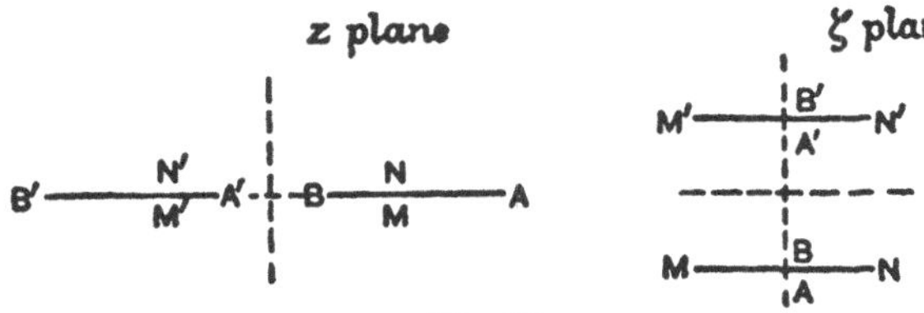

Fig. 97.

and the position of the corresponding points in the two planes
is shown in fig. 97. In particular the extremities of the biplane aerofoils correspond to the stagnation points $z = \pm m$
of the tandem system, and the mid-points of the biplane
aerofoils to the extremities $z = \pm p, \pm q$ of the tandem
aerofoils. The gap h of the biplane system is obtained as the
integral of $\frac{d\zeta}{dz}$ from B to A', and the chord c as twice the
integral of $\frac{d\zeta}{dz}$ from M to A. The respective values are expressed in elliptic integrals in the form

$$h = 2p\left\{E' - \frac{p^2 - m^2}{p^2}K'\right\},$$

$$c = 2p\left\{E(k,\tau) - \frac{m^2}{p^2}F(k,\tau)\right\},$$

where
$$\tau = \sqrt{\frac{p^2 - m^2}{p^2 - q^2}}, \qquad k' = \frac{q}{p},$$

$$E' = \int_0^1 \sqrt{\frac{1 - k'^2 x^2}{1 - x^2}}\,dx,$$

$$K' = \int_0^1 \frac{dx}{\sqrt{(1 - x^2)(1 - k'^2 x^2)}},$$

$$E(k,\tau) = \int_0^\tau \sqrt{\frac{1 - k^2 x^2}{1 - x^2}}\,dx,$$

$$F(k,\tau) = \int_0^\tau \frac{dx}{\sqrt{(1 - x^2)(1 - k^2 x^2)}}.$$

Far from the aerofoils the transformation is $\zeta = -iz$ and
hence to obtain the flow inclined at angle α to the chord of

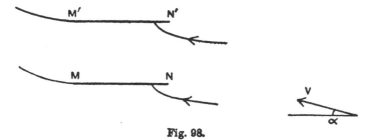

Fig. 98.

the biplane the limiting value of the potential function must
be
$$w = - V (\cos \alpha + i \sin \alpha) \zeta$$
$$= - V (\sin \alpha - i \cos \alpha) z.$$

The general flow in the z plane which will convert into the
desired flow in the ζ plane is now

$$\frac{dw}{dz} = - V \sin \alpha - V \cos \alpha \frac{z^2 - m^2}{\sqrt{(p^2 - z^2)(z^2 - q^2)}}$$
$$+ \frac{1}{\pi} \frac{Cz + C'n}{\sqrt{(p^2 - z^2)(z^2 - q^2)}},$$

where m and n have the values determined previously. In
order to obtain finite velocity at the trailing edges M and M'
of the biplane aerofoils, the numerator over the radical must
be zero at the points M and M' of the tandem system.
Allowing for the change of sign of the radical from M to M',
this condition gives

$$C' = 0, \qquad C = \frac{\pi V \sin \alpha}{m} \sqrt{(p^2 - m^2)(m^2 - q^2)},$$

and the circulation has the same value round both aerofoils.

The resultant force on the biplane system is the lift force

$$L = 2\rho VC = 2\pi\rho V^2 \sin \alpha \frac{\sqrt{(p^2 - m^2)(m^2 - q^2)}}{m},$$

and the lift coefficient of the biplane may therefore be written
in the form
$$C_L = 2B\pi \sin \alpha,$$

where
$$B = \frac{\sqrt{(p^2 - m^2)(m^2 - q^2)}}{mc}.$$

The factor B represents the reduction of the lift coefficient
of an unstaggered biplane compared with that of a monoplane at the same angle of incidence, and values of B are
given in table 17 below. For small angles of incidence the
result may be expressed in the alternative form of the increase
of the angle of incidence required with a biplane to obtain
the same lift coefficient as that of a monoplane. If α is the

angle of incidence of the biplane and α_0 that of the mono-plane, it can easily be shown that

$$\alpha = \alpha_0 + \beta C_L,$$

where

$$\beta = \frac{1-B}{2\pi B}.$$

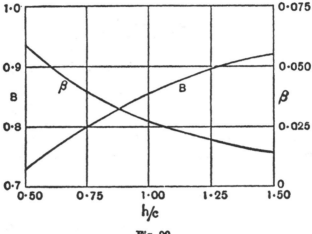

Fig. 99.

The forces on the individual aerofoils of the biplane system can be determined by the same method as in the case of the tandem aerofoils, and it appears that the upper aerofoil experiences a greater lift force than the lower aerofoil.

Table 17.

Correction factors for a biplane.

h/c	B	β
0·50	0·730	0·059
0·75	0·800	0·039
1·00	0·855	0·027
1·25	0·895	0·019
1·50	0·920	0·014

13·13. *The general biplane.*

A staggered biplane system, formed by two equal parallel lines, can be derived from the tandem system by the conformal transformation

$$\frac{d\zeta}{dz} = -\sin\theta + \cos\theta \frac{z^2 - m^2}{\sqrt{(p^2 - z^2)(z^2 - q^2)}},$$

where θ is the angle of stagger of the biplane. More generally also a biplane system with two unequal parallel aerofoils can be obtained by starting with a tandem system with aerofoils of unequal length and by applying a suitable conformal transformation.

The analysis in these more general cases becomes highly complex and in all cases the results obtained apply only to straight line aerofoils. It is useful therefore to develop an approximate method of solving the problem of a biplane system which will give a clearer insight into the mechanism of the interference between the two aerofoils and will provide a method of estimating the effect of the shape of the aerofoil section.

The interference experienced by one aerofoil is due to the distortion of the flow caused by the other aerofoil, and an approximate method of attacking the problem may be based on the conception of replacing the disturbing aerofoil by a point vortex of the correct strength at the centre of pressure of the aerofoil. This method should be satisfactory for large values of the gap-chord ratio and its accuracy in general can be tested by comparing the results which it gives for straight line aerofoils with the accurate results of table 17.

13·14. *Approximate solution.*

The circulation K round the lower aerofoil of the biplane system is assumed to be concentrated at the centre of pressure C. The flow in the neighbourhood

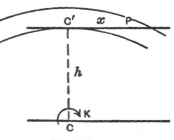

Fig. 100.

of the upper aerofoil due to this circulation K and to the uniform stream V will be curved downwards. At the point P of the upper aerofoil, at distance x behind the point C' which is vertically above the point C, the normal induced velocity due to the circulation K is

$$w = \frac{K}{2\pi} \frac{x}{h^2 + x^2},$$

and the radius of curvature R of the stream lines due to the circulation K and the uniform stream V can be obtained in the form

$$\frac{V}{R} = \frac{dw}{dx} = \frac{K}{2\pi} \frac{h^2 - x^2}{(h^2 + x^2)^2}$$

by equating the alternative forms $\frac{V^2}{R}$ and $V\frac{dw}{dx}$ of the normal acceleration. There is also an increase of the longitudinal velocity in the neighbourhood of the upper aerofoil, but the effect of this increase on the characteristics of the biplane is exactly balanced by an equal decrease for the lower aerofoil, and hence the variation of the longitudinal velocity may be ignored.

The interference experienced by the upper aerofoil will now be represented by the normal induced velocity at the centre of the aerofoil and by the curvature of the stream lines in its neighbourhood, and in developing approximate expressions for this interference the gap-chord ratio of the biplane will be assumed to be large. If θ denotes the centre of pressure coefficient of the lower aerofoil, the normal induced velocity at the centre of the upper aerofoil may be taken to be

$$w_0 = \frac{Kc}{2\pi h^2} (\tfrac{1}{2} - \theta),$$

and the radius of curvature to be

$$\frac{V}{R} = \frac{K}{2\pi h^2}.$$

The lower aerofoil experiences the same interference effects due to the circulation round the upper aerofoil and these expressions may therefore be applied to the biplane as a whole.

The circulation round an aerofoil section is equal to $\frac{1}{2}C_L cV$ and the centre of pressure coefficient can be replaced by the moment coefficient, $C_M = -\theta C_L$. Hence the normal induced velocity becomes

$$w_0 = \frac{V}{8\pi}\left(\frac{c}{h}\right)^2 (C_L + 2C_M),$$

and to obtain the same lift as a monoplane, the angle of incidence of the biplane must be increased by the small angle $\frac{w_0}{V}$.

A further correction is required on account of the curvature of the stream lines. A circular arc aerofoil of radius R, chord c and camber $\gamma_0 = \frac{c}{8R}$ would behave in the curved flow exactly as a straight line aerofoil in a uniform flow, and hence the curvature of the flow is equivalent to a reduction γ_0 of the effective camber of the aerofoils. But for a circular arc aerofoil of camber γ,

$$C_L = 2\pi(\alpha + 2\gamma),$$
$$C_M = -\tfrac{1}{4}C_L - \pi\gamma,$$

and hence to maintain the same lift coefficient the angle of incidence must be increased by $2\gamma_0$ and there will be a corresponding increase of $\pi\gamma_0$ in the moment coefficient. Also the value of γ_0 is

$$\gamma_0 = \frac{c}{8R} = \frac{Kc}{16\pi h^2 V} = \frac{1}{32\pi}\left(\frac{c}{h}\right)^2 C_L.$$

Adding these two corrections, the angle of incidence of the biplane must exceed that of the monoplane by

$$\alpha - \alpha_0 = \frac{1}{16\pi}\left(\frac{c}{h}\right)^2 (3C_L + 4C_M)$$
$$= \frac{1}{8\pi}\left(\frac{c}{h}\right)^2 (C_L + 2C_{M_0}),$$

where C_{M_0} is the value of the moment coefficient at zero lift. Also the slope of the curve of moment coefficient against lift coefficient for the individual aerofoils of the biplane will be

$$\frac{dC_M}{dC_L} = -\frac{1}{4}\left\{1 - \frac{1}{8}\left(\frac{c}{h}\right)^2\right\}.$$

These expressions are only approximations to the true values and have been obtained on the assumption that the gap-chord ratio is large. A comparison with the accurate values for straight line aerofoils, for which C_{M_0} is zero, is obtained by comparing the values of β from table 17 with the approximate expression $\frac{1}{8\pi}\left(\frac{c}{h}\right)^2$.

Table 18.

h/c	β	$\frac{1}{8\pi}(c/h)^2$
0·50	0·059	0·159
0·75	0·039	0·071
1·00	0·027	0·040
1·25	0·019	0·026
1·50	0·014	0·018

The approximate formula gives values which are too large for the ordinary type of biplane system, but it may possibly be used to indicate the effect of a change of aerofoil section. On this basis the angle of incidence of a biplane will be taken to be

$$\alpha = \alpha_0 + \beta\,(C_L + 2C_{M_0})$$
$$= \alpha_0 + \beta\,(\tfrac{3}{2}C_L + 2C_M).$$

This correction from monoplane to biplane is quite important. For a gap-chord ratio of unity the slope of the curve of lift coefficient against angle of incidence is reduced from 6·28 to 5·36, and that of the moment coefficient against lift coefficient from 0·250 to 0·219.

13·2. *Biplane of finite span.*

When the biplane system consists of two aerofoils of finite span, each aerofoil behaves in a manner similar to a monoplane aerofoil and gives rise to a sheet of trailing vortices. The disturbance at any point is then the induced velocity due to the circulation round the aerofoils and to the two sheets of trailing vortices, and the normal induced velocity at any section of one aerofoil exceeds that which would occur for a monoplane aerofoil by the induced velocity of the vortex system of the second aerofoil. In calculating this additional induced velocity, the trailing vortices may be assumed to extend down stream as straight lines in the same manner as for a monoplane aerofoil.

Fig. 101.

The determination of the induced drag of a biplane system is simplified by Munk's equivalence theorem for stagger*, which states that the total induced drag of any multiplane system is unaltered if any of the lifting elements are moved in the direction of motion, provided that the attitude of the elements is adjusted to maintain the same distribution of lift among them. The truth of this theorem follows at once from the fact that the work done by the induced drag is equal to the rate of increase of kinetic energy in the trailing vortex system (see 10·3), and this kinetic energy is unaffected by a geometrical transformation of the type considered in the theorem. By virtue of this theorem any staggered system can be replaced by a corresponding system of zero stagger which will have the same relationship between total lift and drag. The distribution of drag between the aerofoils will be different in the two cases. In a biplane system with forward stagger the upper aerofoil will have less drag and the lower aerofoil will have more drag than in the corresponding biplane system with zero stagger and with the same distribution of lift between the two aerofoils.

In an unstaggered biplane system the induced drag of one aerofoil due to the influence of the trailing vortices of the second aerofoil is equal to the induced drag of the second

* *Isoperimetrische Ausgaben aus der Theorie des Fluges*, Göttingen dissertation, 1918: translated as *NACA*, 121, 1921.

aerofoil due to the trailing vortices of the first aerofoil. For each aerofoil can be divided into a large number of small elements, each of which carries the same small lift force δL, and if P_1 and P_2 are two such elements on the two aerofoils the normal induced velocity at P_1 due to the trailing vortices of P_2 will be equal to the normal induced velocity at P_2 due to the trailing vortices of P_1. Since the lift

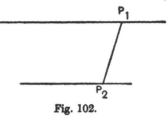

Fig. 102.

forces on the elements are equal, so also will be the induced drag forces due to the element of the other aerofoil. This relationship is true for every pair of elements and by adding the effects of all the elements of the first aerofoil on the element P_2, it appears that the induced drag of the element P_2 due to the trailing vortex system of the first aerofoil is equal to the induced drag of the first aerofoil due to the trailing vortices of the element P_2. Finally by adding the effects of all the elements of the second aerofoil the truth of the theorem is established.

13·21. *The induced drag.*

By virtue of Munk's equivalence theorem for stagger it is sufficient to consider the case of a biplane of zero stagger. Also the lift of each aerofoil will be assumed to be distributed elliptically across the span of each aerofoil, since the load grading curves of most aerofoils approximate closely to this form and the mutual interference of the two aerofoils will be determined with sufficient accuracy by this method.

Let h be the gap of the biplane, s_1 and s_2 the semi-spans of the two aerofoils, L_1 and L_2 the lift forces on the two aerofoils. Then the induced drag forces on the two aerofoils due to their own trailing vortices are respectively

$$D_{11} = \frac{L_1{}^2}{2\pi s_1{}^2 \rho V^2},$$

$$D_{22} = \frac{L_2{}^2}{2\pi s_2{}^2 \rho V^2},$$

and the induced drag of each aerofoil due to the trailing
vortices of the other aerofoil will be of the form

$$D_{12} = \frac{\sigma L_1 L_2}{2\pi s_1 s_2 \rho V^2},$$

where σ is a function of the lengths h, s_1 and s_2.

The values of the coefficient σ can be determined by a
simple graphical method. Fig. 92 shows the normal velocity
at any point above or below a wing with elliptic loading, and
therefore determines the normal induced velocity w_{12} which
occurs at any point of the first aerofoil due to the trailing
vortices of the second aerofoil. The mutual induced drag D_{12}
can then be determined by evaluating the integral

$$D_{12} = \int \frac{w_{12}}{V} dL_1$$

across the span of the first aerofoil. Without any loss of
generality the length s_1 may be assumed to be less than or
equal to the length s_2. The values of the coefficient σ are then

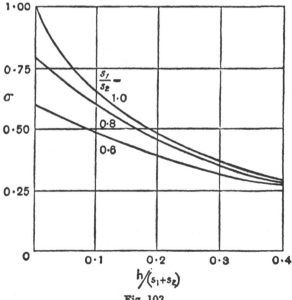

Fig. 103.

determined as a function of the two parameters $\frac{s_1}{s_2}$ and $\frac{h}{s_1 + s_2}$, which are the ratio of the spans of the two aerofoils and the ratio of the gap to the mean span. Numerical values[*] of σ are given in table 19 and fig. 103.

Table 19.

Values of σ.

$h/(s_1 + s_2) =$	0	0·05	0·10	0·15	0·20	0·30	0·40
$s_1/s_2 = 1\cdot0$	1·000	0·780	0·655	0·561	0·485	0·370	0·290
0·8	0·800	0·690	0·600	0·523	0·459	0·355	0·282
0·6	0·600	0·540	0·485	0·437	0·394	0·315	0·255

The induced drag forces of the two aerofoils of the unstaggered biplane are respectively

$$D_1 = D_{11} + D_{12} = \frac{1}{2\pi\rho V^2}\left(\frac{L_1^2}{s_1^2} + \sigma\frac{L_1 L_2}{s_1 s_2}\right),$$

$$D_2 = D_{22} + D_{12} = \frac{1}{2\pi\rho V^2}\left(\frac{L_2^2}{s_2^2} + \sigma\frac{L_1 L_2}{s_1 s_2}\right),$$

and the total induced drag of the biplane is

$$D = \frac{1}{2\pi\rho V^2}\left(\frac{L_1^2}{s_1^2} + 2\sigma\frac{L_1 L_2}{s_1 s_2} + \frac{L_2^2}{s_2^2}\right).$$

For a given total lift $(L_1 + L_2)$ the induced drag is a minimum when

$$\frac{L_1}{L_2} = \frac{s_1(s_1 - \sigma s_2)}{s_2(s_2 - \sigma s_1)},$$

and this equation defines the best distribution of lift between the two aerofoils. The corresponding minimum induced drag is

$$D = \frac{L^2}{2\pi\rho V^2}\frac{1 - \sigma^2}{s_1^2 - 2\sigma s_1 s_2 + s_2^2}.$$

[*] L. Prandtl, "Tragflügeltheorie," *Göttingen Nachrichten*, 1919.

13·22. *Aerofoils of equal span.*

For a biplane with aerofoils of equal span the minimum induced drag occurs when each aerofoil carries the same lift and is given by the equation

$$D = \frac{L^2}{2\pi s^2 \rho V^2} \frac{1 + \sigma}{2},$$

or

$$C_D = \frac{S}{4\pi s^2} C_L^2 \frac{1 + \sigma}{2}.$$

Now for a biplane the aspect ratio A is defined as the value of $\frac{8s^2}{S}$ in order to agree with the definition for a monoplane when the two aerofoils have the same dimensions. The formula for the induced drag coefficient may therefore be written in the alternative form

$$C_D = \frac{1}{\pi A}(1 + \sigma) C_L^2.$$

The induced drag coefficient is increased by the factor $(1 + \sigma)$ above its value for a monoplane aerofoil of the same aspect ratio.

This formula for the induced drag coefficient can be used with good accuracy over a wider range than that to which it strictly applies. The effect of a small change in the distribution of the lift force between the two aerofoils is quite unimportant, for if $L_1 = xL_2$ the general formula for the induced drag of a biplane with aerofoils of equal span may be expressed in the form

$$D = \frac{L^2}{2\pi s^2 \rho V^2} \left\{ \frac{1 + \sigma}{2} + \frac{1 - \sigma}{2} \left(\frac{x - 1}{x + 1} \right)^2 \right\}.$$

Even with the rather extreme values $x = 1·25$ and $\sigma = 0·4$ the additional term represents an increase of only 0·5 % in the induced drag over its minimum value, and it is sufficiently accurate therefore to use the expression for the minimum induced drag in most practical cases.

The formulae have been developed on the basis of elliptic loading across the span of each aerofoil, which should be sufficiently accurate for estimating the mutual interference, but it may be desirable to retain the correcting factor $(1 + \delta)$ for the induced drag of the aerofoils due to their own trailing vortices, which occurs in the theory of monoplane aerofoils (see 11·21). The induced drag coefficient of a biplane with aerofoils of equal span will then be expressed in the form

$$C_D = \frac{1}{\pi A} (1 + \delta + \sigma) C_L{}^2.$$

The total drag coefficient of a biplane exceeds the induced drag coefficient by the profile drag coefficient of the aerofoil section, and hence the total drag coefficient of a biplane with aerofoils of equal span is finally

$$C_D = C_{D_\bullet} + \frac{1}{\pi A} (1 + \delta + \sigma) C_L{}^2,$$

and exceeds that of a monoplane aerofoil of the same aspect ratio by the amount

$$\frac{1}{\pi A} \sigma C_L{}^2.$$

13·23. *Angle of incidence.*

In order to obtain the same lift coefficient from a biplane system as from a monoplane of the same aspect ratio, it is necessary to use a larger angle of incidence, partly on account of the extra induced velocity and partly on account of the direct interference between the aerofoils which occurs in two-dimensional motion. For a biplane with aerofoils of equal span the increase in the drag coefficient over the monoplane aerofoil of the same aspect ratio is $\dfrac{1}{\pi A} \sigma C_L{}^2$, and the corresponding increase in the angle of incidence will be simply $\dfrac{1}{\pi A} \sigma C_L$. Also the correction for the direct interference between the aerofoils has been obtained previously (13·14) in the form $\beta (\tfrac{3}{2} C_L + 2 C_M)$, and hence the total increase in the angle of incidence is

$$\frac{1}{\pi A} \sigma C_L + \beta (\tfrac{3}{2} C_L + 2 C_M).$$

Finally, the angle of incidence of the biplane may be expressed in the form

$$\alpha = \alpha_0 + \frac{1}{\pi A}(1 + \tau + \sigma)\,C_L + \beta\,(\tfrac{3}{2}C_L + 2C_M),$$

where α_0 is the angle of incidence of the aerofoil section in two-dimensional motion which gives the lift coefficient C_L, and τ is the factor which occurs for a monoplane aerofoil in the general case (see 11·22).

13·24. *Summary.*

The characteristics of an unstaggered biplane with aerofoils of equal span are given by the equations

$$\alpha = \alpha_0 + \frac{1}{\pi A}(1 + \tau + \sigma)\,C_L + \beta\,(\tfrac{3}{2}C_L + 2C_M),$$

$$C_D = C_{D_\bullet} + \frac{1}{\pi A}(1 + \delta + \sigma)\,C_L{}^2,$$

where α_0 and $C_{D_\bullet}$ are the characteristics of the aerofoil section in two-dimensional motion corresponding to the lift coefficient C_L, A is the aspect ratio, β and σ are the biplane coefficients given in tables 17 and 19, and τ and δ are the monoplane coefficients of chapter xi depending on the plan form of the aerofoils.

In rough calculations τ and δ may be ignored and C_M may be taken to be equal to $-\tfrac{1}{4}C_L$, so that the expressions for the angle of incidence and drag coefficient of the biplane become approximately

$$\alpha = \alpha_0 + \frac{1}{\pi A}(1 + \sigma)\,C_L + \beta C_L,$$

$$C_D = C_{D_\bullet} + \frac{1}{\pi A}(1 + \sigma)\,C_L{}^2.$$

These formulae for the characteristics of an unstaggered biplane with aerofoils of equal span are valid only over the normal working range of incidence, since the biplane attains a lower maximum value of the lift coefficient than the corresponding monoplane. The reduction of lift at a given angle of incidence in two-dimensional motion is represented by the factor B of table 17, and this factor gives a rough estimate of the reduction which may also be expected in the maximum lift coefficient.

For a biplane formed by two rectangular aerofoils of aspect ratio 6 and with gap-chord ratio unity, the values of the various coefficients are

$$\tau = 0\cdot163, \qquad \delta = 0\cdot046,$$
$$\beta = 0\cdot027, \qquad \sigma = 0\cdot535,$$

and hence, if C_M is taken to be equal to $-\frac{1}{4}C_L$, the formulae give

$$C_D = C_{D_\bullet} + 0\cdot084C_L{}^2,$$
$$\alpha = \alpha_0 + 0\cdot117C_L,$$

or, if the angle of incidence is measured in degrees,

$$\alpha = \alpha_0 + 6\cdot7^\circ C_L.$$

The slope of the curve of lift coefficient against angle of incidence is $3\cdot62$ per radian instead of $6\cdot28$ for the aerofoil section in two-dimensional motion and $4\cdot54$ for a monoplane rectangular aerofoil of the same aspect ratio.

CHAPTER XIV

WIND TUNNEL INTERFERENCE
ON AEROFOILS

14·1. The limited extent of the stream of air in a wind tunnel, whether of open or of closed working section, imposes certain restrictions on the flow past an aerofoil or other body under test, and the determination of the magnitude of this interference is of considerable importance, since it is found that certain corrections must be applied to the aerodynamic characteristics of an aerofoil tested in a wind tunnel before they are applicable to free air conditions. This interference correction is independent of and additional to any correction which may be necessary to allow for the change of scale from a model aerofoil to an actual aeroplane wing.

The theory of the interference has been developed by Prandtl in his second aerofoil paper* by considering the conditions which must be satisfied at the boundary of the stream. The continental wind tunnels usually have an open working section and the condition of constant pressure must be satisfied at the boundary of the stream. British wind tunnels, on the other hand, have a closed working section of square or rectangular cross section, and the boundary condition takes the form that the component of the velocity normal to the tunnel walls must be zero†. This boundary condition can be satisfied analytically by the introduction of a suitable series of images of the model, and the interference experienced by the model is the induced velocity corresponding to the vortex systems of these images. The problem of wind tunnel interference is therefore the choice of the appropriate system of images and the determination of the corresponding induced velocity experienced by the model. The analysis is simplified by the fact that, when the span of the model aerofoil does not exceed three-quarters of

* "Tragflügeltheorie," *Göttingen Nachrichten*, 1919.
† See Note 12 of Appendix.

the breadth of the wind tunnel, it is sufficiently accurate to assume that the lift is distributed uniformly across the span of the aerofoil and that the whole aerofoil experiences the induced velocity which occurs at the centre of the wind tunnel.

14·2. *Tunnel of circular section.*

Consider an aerofoil of semi-span s and of area S in a wind tunnel with closed circular section of radius R. On the assumption of uniform lift distribution across the span of the aerofoil, there will be two trailing vortices only, each of strength K, the circulation round the wing. In the cross sectional plan (fig. 104) these vortices will be situated at A and B on a diameter of the circle representing the boundary of the tunnel and will be at distance s from the centre of the

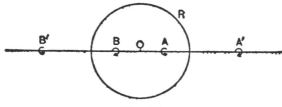

Fig. 104.

circle. The images A' and B' will lie outside the circle on the same diameter at distance $\dfrac{R^2}{s}$ from the centre. The strength of the images will be the same as that of the original vortices but the sense of the circulation will be reversed. This image system depends on the fact that the circle is a stream line for the vortex pairs A, A' and B, B'.

The induced velocity experienced by the aerofoil is the sum of the effects due to the vortices A' and B', and is readily calculated as

$$w = -2\,\frac{K}{4\pi\,.\,OA'} = -\frac{Ks}{2\pi R^2}.$$

The negative sign occurs because the normal velocity w is reckoned positive downwards and the effect of the images is to cause an upward induced velocity at O. By virtue also of the equation for the lift of the aerofoil

$$C_L \cdot \tfrac{1}{2}\rho V^2 S = 2s\rho V K,$$

the result takes the form

$$w = -\frac{C_L S V}{8\pi R^2}.$$

Denote by C the cross sectional area of the tunnel and by ϵ_1 the upward inclination of the stream due to the interference of the boundary or of the images, and then

$$\epsilon_1 = -\frac{w}{V} = \frac{1}{8}\frac{S}{C}C_L.$$

The interference effect is equivalent to an upward inclination of the stream through the angle ϵ_1 and in consequence the lift force is inclined forwards by the same angle, causing a reduction of the drag compared with free air conditions. At the same time the true angle of incidence of the aerofoil will be greater than the inclination of the aerofoil chord to the axis of the tunnel by this same angle ϵ_1. It follows that the corrections which must be applied to the wind tunnel observations on the aerofoil to allow for the constraint of the tunnel walls are of the form

$$\Delta\alpha = \delta\frac{S}{C}C_L,$$

$$\Delta C_D = \delta\frac{S}{C}C_L^2,$$

and, for a tunnel of circular cross section, δ has the numerical value 0.125. The angle of incidence in this formula is, of course, to be taken in circular measure. It will be noticed that the correction is proportional to the lift of the aerofoil and does not depend on the plan form or aspect ratio. The correction can therefore be applied to any wing system, whether monoplane or biplane.

A more detailed analysis of the problem has been given by Prandtl[*], in which he assumes elliptic distribution of lift

---

[*] loc. cit.

across the span of the aerofoil and obtains the value of the coefficient δ in the form

$$\delta = \frac{1}{8}\left\{1 + \frac{3}{16}\left(\frac{s}{R}\right)^{4} + \dots\right\}.$$

Even when the span of the aerofoil is as large as three-quarters of the diameter of the tunnel, the second term represents a correction to the value of δ of only 6 % and is quite negligible in practice.

Prandtl also considers the case of a tunnel of open working section and finds exactly the same form for the correction but with the opposite sign. It appears therefore that the angle of incidence and the drag coefficient, at any definite value of the lift coefficient, are measured too high in a tunnel with open working section and too low in a tunnel with closed working section.

14·3. *Vertical and horizontal boundaries.*

In considering the case of a tunnel of rectangular cross section* it will be assumed that the aerofoil is placed in the centre of the tunnel with its span horizontal. The origin of coordinates will be taken at the centre of the aerofoil with the y axis horizontal to starboard and with the z axis downwards. Before considering the rectangular tunnel, however, it is instructive to consider the effect of the vertical and horizontal boundaries separately.

The system of images for vertical boundaries at distance b apart is illustrated in fig. 105. The images are all identical with the original aerofoil and form an infinite series, uniformly spaced along the y axis at the points $y = \pm\, mb$, where

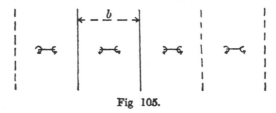

Fig 105.

* H. Glauert, "The interference of wind channel walls on the aerodynamic characteristics of an aerofoil," *RM*, 867, 1923.

m assumes all positive integral values. This system will clearly give zero velocity component normal to the vertical boundaries and so satisfies the conditions of the problem.

Now from 12·22 it appears that the induced velocity far out along the span of an aerofoil can be taken to be

$$w = - \frac{S}{8\pi y^2} C_L V,$$

and to obtain the effect of the system of images, y must be replaced by mb and the summation made for all positive and negative integral values of m. The upward inclination of the stream due to the constraint of the boundary walls is therefore

$$\epsilon_1 = - \frac{w}{V} = \frac{S C_L}{4\pi b^2} \sum_1^\infty \frac{1}{m^2} = \frac{\pi}{24} \frac{S}{b^2} C_L.$$

The case of horizontal boundary walls above and below the aerofoil is treated on similar lines. An infinite series of images is again required, situated on the z axis at the points $z = \pm nh$, but these images are of alternate sign, being positive like the original aerofoil when n is even, and negative when n is odd. The induced velocity far out along the z axis of an aerofoil is (from 12·22)

$$w = \frac{S}{8\pi z^2} C_L V,$$

and hence the effect of the whole system of images is

$$\epsilon_1 = - \frac{w}{V} = - \frac{S C_L}{4\pi h^2} \sum_1^\infty \frac{(-1)^n}{n^2} = \frac{\pi}{48} \frac{S}{h^2} C_L.$$

A comparison of this result with the previous one shows that the lateral vertical boundaries exert a greater interference on the aerofoil than the horizontal boundaries parallel to the span. In each case the effect of the interference is to reduce the angle of incidence and drag coefficient in a wind tunnel compared with free air conditions.

14·4. *Tunnel of rectangular section.*

For a rectangular tunnel of breadth b and height h, a doubly infinite series of images is required, situated at the points ($y = mb$, $z = nh$), where m and n assume all positive and negative integral values except the pair $(0, 0)$. The lift of the image is positive when n is even, and negative when n is odd.

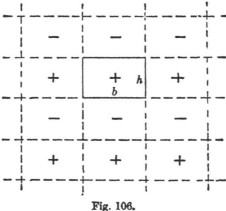

Fig. 106.

The induced velocity at the point (y, z) far from an aerofoil is (from 12·22)

$$w = -\frac{1}{8\pi}\frac{y^2 - z^2}{(y^2 + z^2)^2}SC_LV,$$

and hence the effect of the whole system of images is to cause an upward inclination of the stream

$$\epsilon_1 = -\frac{w}{V} = \frac{SC_L}{8\pi}\sum_{-\infty}^{\infty}\sum_{-\infty}^{\infty}(-1)^n\frac{m^2b^2 - n^2h^2}{(m^2b^2 + n^2h^2)^2}$$

$$= \frac{SC_L}{8\pi b^2}\sum_{-\infty}^{\infty}\sum_{-\infty}^{\infty}(-1)^n\frac{m^2 - \lambda^2n^2}{(m^2 + \lambda^2n^2)^2},$$

where $h = \lambda b$. The double summation extends over all the images, i.e. over all integral values of m and n except the pair $(0, 0)$. No simple form has been found for the sum of this series, but the following method of reduction can be employed. Starting from the expansion*

$$\cot z = \frac{1}{z} + 2z\sum_{1}^{\infty}\frac{1}{z^2 - m^2\pi^2},$$

* Hobson, *Plane Trigonometry*, p. 334.

put
$$z = i\lambda\pi x$$

and obtain
$$\coth \lambda\pi x = \frac{1}{\lambda\pi x} + \frac{2\lambda x}{\pi} \sum_1^\infty \frac{1}{m^2 + \lambda^2 x^2}.$$

Then directly and by differentiation with respect to x

$$\sum_1^\infty \frac{1}{m^2 + \lambda^2 x^2} = -\frac{1}{2\lambda^2 x^2} + \frac{\pi}{2\lambda x} \coth \lambda\pi x,$$

$$\sum_1^\infty \frac{-2\lambda^2 x^2}{(m^2 + \lambda^2 x^2)^2} = \frac{1}{\lambda^2 x^2} - \frac{\pi}{2\lambda x} \coth \lambda\pi x - \frac{\pi^2}{2} \operatorname{cosech}^2 \lambda\pi x,$$

and by addition

$$\sum_1^\infty \frac{m^2 - \lambda^2 x^2}{(m^2 + \lambda^2 x^2)^2} = \frac{1}{2\lambda^2 x^2} - \frac{\pi^2}{2} \operatorname{cosech}^2 \lambda\pi x.$$

This result leads to the summation

$$\sum_1^\infty \sum_1^\infty (-1)^n \frac{m^2 - \lambda^2 n^2}{(m^2 + \lambda^2 n^2)^2} = \frac{1}{2\lambda^2} \sum_1^\infty \frac{(-1)^n}{n^2} - \frac{\pi^2}{2} \sum_1^\infty (-1)^n \operatorname{cosech}^2 \lambda\pi n$$

$$= -\frac{\pi^2}{24\lambda^2} - 2\pi^2 \sum_1^\infty \sum_1^\infty (-1)^n p e^{-2\lambda\pi np}$$

$$= -\frac{\pi^2}{24\lambda^2} + 2\pi^2 \sum_1^\infty \frac{p}{1 + e^{2\lambda\pi p}},$$

and then finally

$$\sum_{-\infty}^\infty \sum_{-\infty}^\infty (-1)^n \frac{m^2 - \lambda^2 n^2}{(m^2 + \lambda^2 n^2)^2} = 4 \sum_1^\infty \sum_1^\infty (-1)^n \frac{m^2 - \lambda^2 n^2}{(m^2 + \lambda^2 n^2)^2}$$

$$+ 2 \sum_1^\infty \frac{1}{m^2} - \frac{2}{\lambda^2} \sum_1^\infty \frac{(-1)^n}{n^2}$$

$$= \frac{\pi^2}{3} + 8\pi^2 \sum_1^\infty \frac{p}{1 + e^{2\lambda\pi p}},$$

from which numerical results can be rapidly obtained, since it is sufficiently accurate to retain only the first term of the last exponential series.

Expressing the angle of deviation ϵ_1 of the stream in the form

$$\epsilon_1 = \delta \frac{S}{C} C_L,$$

where C is the cross sectional area of the wind tunnel, the following numerical values are obtained:

Rectangular Tunnels						
h/b	1/4	1/2	$1/\sqrt{2}$	1	2	4
δ	0·262	0·137	0·119	0·137	0·262	0·524

These numerical values show the curious result that, for the range of values considered, the interference is unaltered if a tunnel of breadth b and height h is replaced by one of breadth $\sqrt{2}h$ and height $\dfrac{b}{\sqrt{2}}$. The best ratio of breadth to height, for a given cross sectional area, is $\sqrt{2}$ and the interference is then slightly less than that in a tunnel of circular section for which δ has the value 0·125.

14·5. *Downwash and tailsetting.*

The preceding analysis relates to the interference experienced by an aerofoil or system of wings due to the constraint of the tunnel walls, and leads to corrections which must be applied to the angle of incidence and drag coefficient measured in a wind tunnel. The interference was found to be of the form of an upwash angle ϵ_1 and this interference will be increased* by an additional angle ϵ_2 in the neighbourhood of the tailplane of a model aeroplane. In consequence the downwash angle ϵ and the tailsetting α_T to trim the aeroplane will be measured smaller in a wind tunnel than in free air and will require the corrections

$$\Delta\epsilon = \epsilon_1 + \epsilon_2,$$
$$\Delta\alpha_T = \epsilon_2.$$

To calculate the angle ϵ_2 it is necessary to determine the induced velocity due to the system of images in the neighbourhood of the tailplane of the model. Consider first the effect of a single aerofoil with uniform lift distribution across

* Glauert and Hartshorn, "The interference of wind channel walls on the downwash angle and tailsetting," *RM*, 947, 1924.

the span at the general point (x, y, z), where x is measured downstream, y to starboard and z downwards. The complete expression for the normal induced velocity is (from 12·2)

$$\frac{4\pi w}{K} = \frac{x}{x^2 + z^2} \left\{ \frac{y + s}{\sqrt{(y + s)^2 + x^2 + z^2}} - \frac{y - s}{\sqrt{(y - s)^2 + x^2 + z^2}} \right\}$$
$$+ \frac{y + s}{(y + s)^2 + z^2} \left\{ 1 + \frac{x}{\sqrt{(y + s)^2 + x^2 + z^2}} \right\}$$
$$- \frac{y - s}{(y - s)^2 + z^2} \left\{ 1 + \frac{x}{\sqrt{(y - s)^2 + x^2 + z^2}} \right\},$$

but this expression can be simplified by the assumption that x is of the same order as s and that $\sqrt{y^2 + z^2}$ is large compared with s. To the first order of approximation, as used in the preceding analysis, the value of w then becomes

$$\frac{w}{V} = - \frac{1}{8\pi} \frac{y^2 - z^2}{(y^2 + z^2)^2} SC_L - \frac{1}{8\pi} \frac{x(y^2 - 2z^2)}{(y^2 + z^2)^{\frac{5}{2}}} SC_L,$$

where K has been replaced by $\frac{S}{4s} C_L V$.

The first of these terms is independent of x and is the expression which has been used previously to calculate the value of ϵ_1. The second term represents the additional interference ϵ_2 experienced by the tailplane of a model aeroplane, and hence the value of ϵ_2 for a tunnel of rectangular section must be calculated as

$$\epsilon_2 = \frac{xSC_L}{8\pi b^3} \sum_{-\infty}^{\infty} \sum_{-\infty}^{\infty} (-1)^n \frac{m^2 - 2\lambda^2 n^2}{(m^2 + \lambda^2 n^2)^{\frac{5}{2}}},$$

where $h = \lambda b$ as before. This result can be written in the form

$$\epsilon_2 = \delta' \frac{xS}{hC} C_L,$$

and the following numerical values have been calculated for the coefficient δ':

$$\lambda = \tfrac{1}{2}, \quad \delta' = 0·293,$$
$$\lambda = 1, \quad \delta' = 0·240.$$

14·6. *Summary.*

The aerodynamic characteristics of a model aeroplane obtained from tests in a wind tunnel with closed working section require the following corrections to allow for the interference of the tunnel walls:

Angle of incidence $\quad \Delta\alpha = \epsilon_1$,

Tailsetting $\qquad\qquad \Delta\alpha_T = \epsilon_2$,

Downwash angle $\quad\; \Delta\epsilon = \epsilon_1 + \epsilon_2$,

Drag coefficient $\quad\; \Delta C_D = \epsilon_1 C_L$,

where ϵ_1 and ϵ_2 are defined by the equations

$$\epsilon_1 = \delta \frac{S}{C} C_L,$$

$$\epsilon_2 = \delta' \frac{xS}{hC} C_L,$$

and S = total wing area of the model,

$\quad x$ = distance of tailplane from centre of gravity,

$\quad C$ = cross sectional area of the tunnel,

$\quad h$ = height of the tunnel, normal to the wing span.

All angles are to be taken in circular measure and the coefficients δ and δ' have the following values in typical cases:

Wind Tunnel	δ	δ'
Circular	0·125	—
Square	0·137	0·240
Rectangular ($b = 2h$)	0·137	0·293

CHAPTER XV

THE AIRSCREW: MOMENTUM THEORY

15·1. An airscrew normally consists of a number of equally spaced identical radial arms, and the section of a blade at any radial distance r has the form of an aerofoil section whose chord is set at an angle θ to the plane of rotation. The blade angle θ and the camber of the aerofoil section decrease outwards along the blade. If the airscrew moved through the air as through a solid medium, the advance per revolution would be $2\pi r \tan \theta$ and this quantity would define the pitch of the screw. Actually this quantity will not have the same value for all radial elements of the blade and so it is customary to define as the geometrical pitch of the airscrew the value of $2\pi r \tan \theta$ at a radial distance of 70 per cent. of the tip radius. An airscrew rotates in a yielding fluid and in consequence the advance per revolution is not the same as the geometrical pitch and may in fact assume any value. The value of the advance per revolution for which the thrust of the airscrew vanishes is called the experimental mean pitch, and in many respects the characteristics of an airscrew are defined by the ratio of the experimental mean pitch to the diameter.

An ordinary propulsive airscrew experiences a torque or couple resisting its rotation and gives a thrust along its axis. The thrust T and the torque Q are expressed as functions of the axial velocity V, the number of revolutions in unit time n (or the angular velocity Ω) and the diameter D. The state of operation of the airscrew is defined by the advance per revolution, but it is preferable to express this parameter in the non-dimensional form

$$J = \frac{V}{nD}.$$

The standard British non-dimensional coefficients for the thrust and torque of an airscrew are

$$k_T = \frac{T}{\rho n^2 D^4} \quad \text{and} \quad k_Q = \frac{Q}{\rho n^2 D^5},$$

but it is convenient at times to use the alternative coefficients

$$T_C = \frac{T}{\rho V^2 D^2} \quad \text{and} \quad Q_C = \frac{Q}{\rho V^2 D^3}.$$

Other forms of these coefficients are used by continental writers and a variety of forms can be obtained by using the angular velocity Ω instead of n, the disc area $\frac{\pi}{4} D^2$ instead of D^2, and the dynamic pressure $\frac{1}{2}\rho V^2$ instead of ρV^2. These coefficients are all simple numerical multiples of those defined above and each form has its own merits in particular cases.

Airscrews are used for a variety of purposes, of which the following may be mentioned:

(1) *Propeller*. An airscrew used for propulsion, as on an aircraft, and designed to give a high thrust power TV for a given torque power ΩQ.

(2) *Windmill*. An airscrew used to obtain torque power from its axial motion relative to the air. A distinction must be drawn between a windmill mounted on an aeroplane, where the drag is of importance and the axial velocity is high, and one fixed on the ground, where the drag is unimportant and the axial velocity is low.

(3) *Fan*. An airscrew used to obtain a current of air.

(4) *Anemometer*. An airscrew used to determine the relative axial velocity by means of the rate of rotation.

The theory of the behaviour of an airscrew follows the same lines, whatever the purpose for which it is intended. The design will vary, however, and apart from aerodynamic considerations, limitations are also imposed by considerations of strength and size. Other types of instrument may serve the same purpose as an airscrew and, in particular, hemispherical cups mounted at the end of radial arms are used both as windmills and as anemometers, but these instruments form a separate class distinct from airscrews.

When an airscrew has a large diameter or high rate of rotation, the tip velocity may rise to the same order of magnitude as the velocity of sound and the compressibility of the air will then modify the forces experienced by the blade elements. This effect does not become of importance until the tip velocity ($\pi n D$) exceeds 800 f.p.s. and in developing the theory of an airscrew it will be assumed that the effect of the compressibility of the air may be neglected. No theory has been developed, as yet, which takes account of the compressibility effect and the modification to the characteristics of an airscrew due to high tip speed must be estimated from special experimental investigations.

15·2. *Simple momentum theory.*

A simple method of considering the operation of an airscrew, based on the work of Rankine and Froude, depends on a consideration of the momentum and kinetic energy of the system. The airscrew is assumed to have a large number of blades, so that it becomes effectively a circular disc, and it is further assumed that the thrust is uniformly distributed over this disc. The rotation of the slipstream due to the action of the torque is ignored* and the axial velocity of the fluid must be continuous in passing through the airscrew disc in order to maintain continuity of the flow. On the other hand, the pressure of the fluid receives a sudden increment p', equal to the thrust on unit area of the disc, and a slipstream of increased axial velocity is formed behind the airscrew. The term "actuator disc" has been given to this simplified conception of an airscrew and a number of interesting results can be derived by considering the momentum and energy of the slipstream.

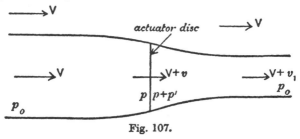

Fig. 107.

* An extension of the momentum theory of an airscrew including the rotation has been given by A. Betz, "Eine Erweiterung der Schraubenstrahl-Theorie," *ZFM*, 1920.

Consider an actuator disc in a stream of velocity V, for which the general type of flow will be as shown in fig. 107. On approaching the disc the axial velocity rises to $(V + v)$ and the pressure falls from p_0 to p. The axial velocity is constant in passing through the disc but rises to $(V + v_1)$ in the final slipstream, and the pressure rises to $(p + p')$ immediately behind the disc and then falls to its original value p_0. The whole flow is regarded as irrotational except for the discontinuity of pressure in passing through the airscrew disc, and hence it is legitimate to apply Bernoulli's equation to the motion before and behind the disc separately. The total pressure head in these two regions has the values

$$H_0 = p_0 + \tfrac{1}{2}\rho V^2 = p + \tfrac{1}{2}\rho (V + v)^2,$$
$$H_1 = p_0 + \tfrac{1}{2}\rho (V + v_1)^2 = p + p' + \tfrac{1}{2}\rho (V + v)^2,$$

and hence
$$p' = H_1 - H_0 = \rho (V + \tfrac{1}{2}v_1) v_1.$$

Also by considering the rate of increase of axial momentum it appears that the thrust is

$$T = A\rho (V + v) v_1,$$

where A is the area of the actuator disc, and since p' is the thrust on unit area of the disc

$$p' = \rho (V + v) v_1.$$

By comparing the two expressions for p', it follows that

$$v = \tfrac{1}{2}v_1.$$

Thus half the added velocity in the slipstream occurs before the airscrew and half behind it, and the thrust of the airscrew becomes
$$T = 2A\rho (V + v) v.$$

The increase of kinetic energy of the fluid in unit time is

$$E = \tfrac{1}{2}A\rho (V + v) \{(V + v_1)^2 - V^2\}$$
$$= 2A\rho (V + v)^2 v$$
$$= T (V + v),$$

which is the work done on the fluid by the thrust of the airscrew. Also if Ω is the angular velocity and Q is the torque of the airscrew, the total work done on the airscrew is ΩQ, and it follows that
$$\Omega Q = T (V + v).$$

Consider next the case when the general mass of the fluid is at rest and the airscrew is advancing with velocity V. The relationship between the thrust and the velocity is unaltered but the work done by the thrust is now TV on the airscrew and Tv on the fluid, and this second term is equal to the rate of increase of kinetic energy of the fluid

$$E = \tfrac{1}{2}A\rho \, (V + v) \, v_1{}^2$$
$$= 2A\rho \, (V + v) \, v^2$$
$$= Tv.$$

15·21. *Ideal efficiency.*

The efficiency of propulsion of the airscrew, defined as the ratio of the useful work to the total work, is

$$\eta = \frac{TV}{\Omega Q} = \frac{V}{V + v},$$

and it is customary to write $v = aV$, so that the efficiency becomes

$$\eta = \frac{1}{1 + a}.$$

This expression represents the ideal efficiency of an airscrew which is never fully realised in practice. The ideal efficiency has been obtained on the assumption that the only loss of energy is represented by the kinetic energy of the axial velocity in the slipstream, but the following additional sources of loss of energy exist:

(1) frictional drag of the airscrew blades,

(2) kinetic energy of the rotation of the slipstream,

(3) periodicity of flow and loss of thrust towards the blade tips, so that the thrust is not uniformly distributed over the airscrew disc.

The most important of these additional effects is usually the frictional drag of the airscrew blades, and under ordinary working conditions of an airscrew the actual efficiency is about 85 % of the ideal efficiency. An examination of the ideal efficiency is therefore a useful guide to the actual efficiency which may be anticipated from an airscrew.

Since
$$T = 2A\rho V^2 (1 + a)\, a,$$

$$T_C = \frac{T}{\rho V^2 D^2} = \frac{\pi}{2}(1 + a)\, a,$$

and
$$1 + 2a = \sqrt{1 + \frac{8}{\pi}\, T_C}$$

$$= \sqrt{1 + \frac{8}{\pi}\frac{k_T}{J^2}},$$

where that root of the quadratic equation has been taken which makes a vanish with the thrust. When V is zero, a becomes infinite but the velocity of flow through the airscrew has the finite value

$$\frac{v}{nD} = \frac{aV}{nD} = \sqrt{\frac{2}{\pi}\, k_T}.$$

Now suppose that power P is put into an airscrew of diameter D. Equating ηP to the work done by the thrust

$$\eta P = TV = \frac{\pi}{2} D^2 \rho V^3 (1 + a)\, a,$$

or
$$\frac{1 - \eta}{\eta^3} = \frac{2}{\pi}\frac{P}{\rho V^3 D^2},$$

which is an equation to determine the ideal efficiency in terms of the power, airscrew diameter and speed. The power must be expressed in units which are consistent with those of the other quantities, and in British Engineering units the power is to be expressed in ft. lb. per sec. The relationship between the efficiency and the power coefficient is given in table 20 and fig. 108. The efficiency falls rapidly as the power coefficient increases and this fall represents the loss entailed by putting a large power through an airscrew of small diameter.

Table 20.

Power and Ideal Efficiency.

η	$P/\rho V^3 D^2$	η	$P/\rho V^3 D^2$	η	$P/\rho V^3 D^2$
100	0	90	0·216	80	0·614
$97\frac{1}{2}$	0·042	$87\frac{1}{2}$	0·294	$77\frac{1}{2}$	0·759
95	0·092	85	0·384	75	0·932
$92\frac{1}{2}$	0·149	$82\frac{1}{2}$	0·490	$72\frac{1}{2}$	1·133

The power given by an engine is a simple function of the rate of revolution for given conditions of pressure and

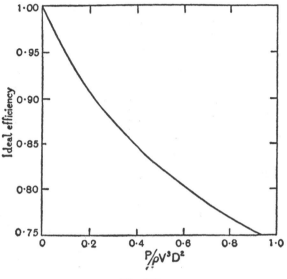

Fig. 108.

temperature, and so it is convenient for some purposes to write the expression for the ideal efficiency in the modified form

$$\frac{1-\eta}{\eta^3} = \frac{2}{\pi}\frac{1}{J^3}\frac{P}{\rho n^3 D^5},$$

where n is the number of the airscrew revolutions per sec., which differ from those of the engine when gearing is employed. It is now possible to draw the curve of η against J for any definite value of $\frac{P}{\rho n^3 D^5}$, and this curve will represent the ideal efficiency of an airscrew which is adjusted to run at a constant rate of revolution by altering the pitch of the blades. Typical curves are given in fig. 109 which show the variation of the ideal efficiency with the rate of advance of the airscrew. The ideal efficiency increases with J, rapidly

at first and then more slowly, and approaches unity as a
limit.

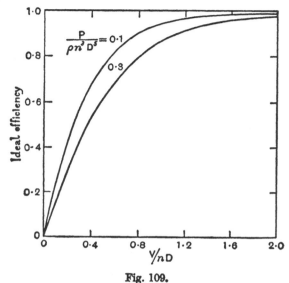

Fig. 109.

15·22. *Windmills*.

The simple momentum theory can also be applied to the
problem of a windmill designed to take power from its motion
relative to the fluid. The windmill experiences a negative
thrust or drag and the fluid in the slipstream is retarded, so
that it is convenient to write $v' = -v$. The equation connecting this velocity with the drag R is

$$R = 2A\rho \ (V - v') \ v'.$$

Consider first the case of a windmill on an aeroplane. The
energy put into the air by the windmill in unit time is

$$E = 2A\rho \ (V - v') \ v'^2,$$

while the work done by the aeroplane on the windmill is RV.
It follows that the power which can be taken from the windmill is

$$P = \Omega Q = RV - E = 2A\rho \ (V - v')^2 \ v',$$

and the efficiency of the windmill can be defined as

$$\eta' = \frac{\Omega Q}{R\dot{V}} = \frac{V - v'}{V}.$$

This is exactly the inverse of the efficiency of propulsion of an airscrew.

The relationship between the efficiency and power output of the windmill is

$$\eta'^2 (1 - \eta') = \frac{2}{\pi} \frac{P}{\rho V^3 D^2}.$$

For a given speed and diameter, the power is a maximum when $\eta' = \frac{2}{3}$ and $v' = \frac{1}{3}V$, and the maximum power has the value

$$P \text{ (max)} = \frac{2\pi}{27} \rho V^3 D^2 = 0.232 \rho V^3 D^2.$$

The case of a windmill on the ground exhibits different features since the drag is now unimportant. The power given by the windmill has the same value as in the previous case, but the efficiency requires a new definition. If the windmill created no disturbance of the flow, the energy of the air passing through the windmill disc in unit time would be

$$E = \tfrac{1}{2} A \rho V^3,$$

and the efficiency may suitably be defined as the ratio of the power given by the windmill to this quantity. The efficiency then is

$$\eta'' = \frac{4 (V - v')^2 v'}{V^3},$$

which has the maximum value

$$\eta'' \text{ (max)} = \tfrac{16}{27} = 0.593.$$

CHAPTER XVI

THE AIRSCREW: BLADE ELEMENT THEORY

16·1. In order to obtain a more detailed knowledge of the behaviour of an airscrew than is given by the simple momentum theory, it is necessary to investigate the forces experienced by the airscrew blades and to regard each element of a blade as an aerofoil element moving in its appropriate manner. It is convenient, in developing the theory, to consider an ordinary propulsive airscrew under ordinary working conditions. The conditions for other types of airscrew and for other working conditions can then be examined as modifications of the main theory.

The airscrew will be assumed to have an angular velocity Ω about its axis and to be placed in a uniform stream of velocity V parallel to the axis of rotation. The sections of the blades of the airscrew have the form of aerofoil sections and the lift force experienced by a blade element in its motion relative to the fluid must be associated with circulation of the flow round the blade. Owing to the variation of this circulation along the blade from root to tip, trailing vortices will spring from the blade and pass downstream with the fluid in approximately helical paths. These vortices are concentrated mainly at the root and tips of the blades and so the slipstream of the airscrew consists of a region of fluid in rotation with a strong concentration of vorticity on the axis and on the boundary of the slipstream. From, the analogy of the general aerofoil theory it follows that the blade element should be regarded as an aerofoil in two dimensional motion, subject to the interference flow represented by the velocity field of the system of trailing vortices. The exact evaluation of the interference flow is of great complexity owing to the periodicity of the flow, but for most purposes it is sufficiently accurate to replace the actual periodical flow by its mean value. This step is equivalent to assuming that, for the purpose of estimating the interference flow due to the system of trailing vortices, the thrust and torque carried on the finite number of blade elements at any radial distance from the axis may be replaced by a uniform distribution of thrust and torque over the whole circumference of the circle of the same radius.

In developing the theory it will also be assumed that the angular velocity of the airscrew does not become so great that the rotational velocity of the blade tips approaches too closely to the velocity of sound. Little is known of the effect of the compressibility of the air on the characteristics of an aerofoil moving with high velocity and further progress, both in theory and in experiment, is necessary before the theory of the airscrew can be modified to take account of this effect.

16·11. In discussing the nature of the flow past an airscrew it is convenient to use the following terms:

Inflow. The flow immediately in front of the airscrew.

Outflow. The flow immediately behind the airscrew.

Wake. The flow in the slipstream far behind the airscrew.

Interference flow. The velocity field of the system of trailing vortices which acts as an interference on the blade elements.

Considering first the rotational motion, it is evident that the torque of the airscrew will cause a rotation about the axis of the flow in the slipstream and that no rotation of this nature can occur in front of the airscrew* or outside the boundary of the slipstream. This rotational motion is to be ascribed partly to the system of trailing vortices and partly to the circulation round the blades. Due to the trailing vortices the flow in the plane of the airscrew will have an angular velocity ω in the same sense as the rotation of the airscrew, and the circulation round the airscrew blades will cause equal and opposite angular velocities of the inflow and outflow. The sum of these two components must be zero in

* This result follows at once from the general theorem of 4·31. See also G. I. Taylor, "The rotational inflow factor in propeller theory," *RM*, 765, 1921.

the inflow, since no rotation is possible until the flow has reached the vortex system generated by the airscrew. Hence it follows that the angular velocity of the outflow is 2ω and that the interference flow, which is due solely to the system of trailing vortices, will have the angular velocity ω.

The angular momentum of the outflow is closely related to the torque of the airscrew. Consider the blade elements dr at radial distance r from the axis, let dQ be the torque of these elements and let u be the axial velocity through the airscrew annulus. Then, by equating the torque to the rate of increase of angular momentum, it appears that

$$dQ = 2\pi r \, dr \, . \rho u \, . \, 2\omega r^2,$$

or
$$\frac{dQ}{dr} = 4\pi r^3 \rho \, V\Omega \, (1 + a) \, a',$$

where
$$u = V \, (1 + a),$$
$$\omega = \Omega a'.$$

The quantities a and a' represent the interference flow and are called the interference factors for the axial and rotational motion respectively.

16·12. The axial velocity must be continuous in passing through the airscrew disc and will have the same value u in the inflow and outflow. The increment of u above the undisturbed stream velocity V is due wholly to the system of trailing vortices and the axial interference velocity is $(u - V)$ or aV. In estimating the magnitude of this axial interference flow, it is assumed that the trailing vortices pass downstream in regular helices. This assumption is equivalent to neglecting the contraction of the slipstream diameter which actually occurs, and may require modification when the interference factor a ceases to be small. The induced velocity of this ideal vortex cylinder at a point of the wake will be double the induced velocity at the airscrew disc which is the end of the cylinder, and hence the axial velocity in the wake is $V \, (1 + 2a)$. This result is in agreement with the conclusion drawn from the simple momentum theory.

The axial momentum equation for the blade elements can now be written down directly as

$$\frac{dT}{dr} = 4\pi r\rho V^2 (1 + a)\, a.$$

This equation is not exact. It is based on the assumption of no contraction of the slipstream in estimating the interference velocity and it also neglects the fact that reduced pressure occurs in the wake owing to the rotational motion. The error introduced by these simplifications appears to be negligible for a propulsive airscrew under ordinary working conditions, but it may be necessary to replace this momentum equation by a more accurate relationship in certain special cases, as for example when an airscrew is rotating at zero rate of advance.

16·13.   A consideration of the system of trailing vortices leads to the interesting conclusion that the interference flow experienced by the blade elements at radial distance r from the axis depends solely on the forces experienced by these elements, and is not influenced by the blade elements at greater or less radial distance. Consider the action of the blade elements dr at radial distance r when the remainder of each airscrew blade is inoperative. The trailing vortices which spring from the ends of the element lie on the surfaces of two circular cylinders of radius r and $r + dr$ respectively, and the vorticity may be resolved into two components, one having its axis parallel to the airscrew axis and the other being circumferential and similar to a succession of vortex rings. The first component of the vorticity acts as the roller bearings between the rotating shell of air bounded by the cylindrical surfaces and the general mass of air. Now the general mass of air cannot acquire any circulation about the axis and hence the rotation due to the torque of the blade elements is confined to the region between the two cylindrical surfaces. Hence also the rotational interference due to the vortex system is experienced only by those blade elements which gave rise to the vorticity.

A similar argument can be applied to the second component

of the vorticity and thus the independence of the blade elements at different radial distances from the axis of the airscrew is established. This theoretical result is of great importance and has been confirmed for the principal working sections of an airscrew blade by certain special experiments*. Towards the tips of the airscrew blades the conditions may be modified by the radial flow of the air which is neglected in developing the theory.

16·2. Consider next the aerodynamic forces experienced by the blade element at radial distance r. The blade element

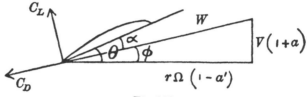

Fig. 110.

is subject to an axial velocity $V(1+a)$ and a rotational velocity $r\Omega(1-a')$, so that the resultant velocity W is inclined at angle ϕ to the plane of rotation, where

$$\tan\phi = \frac{V}{r\Omega}\cdot\frac{1+a}{1-a'}.$$

If θ is the blade angle, the element will work at an angle of incidence $\alpha = \theta - \phi$ and will give the corresponding lift and drag coefficients, C_L and C_D, appropriate to the aerofoil section in two-dimensional motion. The components of these force coefficients, resolved in the direction of the thrust and torque, are respectively

$$\lambda_1 = C_L\cos\phi - C_D\sin\phi,$$
$$\lambda_2 = C_L\sin\phi + C_D\cos\phi,$$

and the elements of thrust and torque given by the blade element of area $c\,dr$ are

$$dT = \lambda_1\tfrac{1}{2}\rho W^2 c\,dr,$$
$$dQ = \lambda_2\tfrac{1}{2}\rho W^2 c r\,dr.$$

* Lock, Bateman and Townend, "Experiments to verify the independence of the elements of an airscrew blade," *RM*, 953, 1924.

These expressions are multiplied by N, the number of blades, to obtain the elements of thrust and torque for the whole airscrew, and in place of the chord a non-dimensional quantity s is introduced, defined by the equation

$$s = \frac{Nc}{2\pi r}.$$

This quantity s represents the ratio of the area of the blade elements to the area of the annulus at the radial distance r, and may be termed the *solidity* of the blade element.

The elements of thrust and torque of the airscrew can now be expressed in the following forms:

$$\frac{dT}{dr} = \pi s r \rho V^2 (1+a)^2 \lambda_1 \operatorname{cosec}^2 \phi$$

$$= \pi s r^3 \rho \Omega^2 (1-a')^2 \lambda_1 \sec^2 \phi,$$

$$\frac{dQ}{dr} = \pi s r^2 \rho V^2 (1+a)^2 \lambda_2 \operatorname{cosec}^2 \phi$$

$$= \pi s r^4 \rho \Omega^2 (1-a')^2 \lambda_2 \sec^2 \phi.$$

Expressions for the elements of thrust and torque have been obtained earlier in the chapter by considering the axial and rotational momenta, and by equating the alternative forms the following equations are obtained for the axial and rotational interference factors:

$$\frac{a}{1+a} = \frac{\frac{1}{2}s\lambda_1}{1 - \cos 2\phi},$$

$$\frac{a'}{1-a'} = \frac{\frac{1}{2}s\lambda_2}{\sin 2\phi}.$$

Finally, the rate of advance of the airscrew is given by the equation

$$J = \frac{V}{nD} = \pi \frac{r}{R} \cdot \frac{V}{r\Omega} = \pi \frac{r}{R} \frac{1-a'}{1+a} \tan \phi,$$

and the elements of thrust and torque can be expressed in the non-dimensional forms

$$R\frac{dk_T}{dr} = \frac{1}{4}\pi^3 \left(\frac{r}{R}\right)^3 s (1-a')^2 \lambda_1 \sec^2 \phi,$$

$$R\frac{dk_Q}{dr} = \frac{1}{8}\pi^3 \left(\frac{r}{R}\right)^4 s (1-a')^2 \lambda_2 \sec^2 \phi.$$

16·21. The method of calculating the characteristics of an airscrew is to choose a number of elements along the blade, for each of which the values of $\frac{r}{R}$, s, θ and the aerofoil characteristics (α, C_L, C_D) are known. Starting with a series of values of α for each element, it is possible to calculate in turn the corresponding values of a, a', J, dk_T and dk_Q.

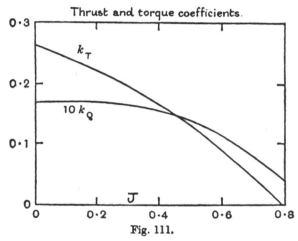

Fig. 111.

Details of the calculation for a typical blade element are given in table 21. Curves of dk_T and dk_Q against J are then

Table 21.

Calculations for a blade element.

$r/R = 0.70$, $s = 0.10$, $\theta = 24°$.

α	ϕ	λ_1	λ_2	a	a'	J	$R\,(dk_T/dr)$	$R\,(dk_Q/dr)$
4°	20°	−·012	·036	−·003	·003	·80	−·004	·0038
6	18	·204	·088	·056	·007	67	·059	0089
8	16	·410	·132	156	·012	·54	·115	·0130
10	14	·610	·164	·353	·017	·40	·166	·0156
12	12	·780	·178	·820	·021	·25	·207	·0166
14	10	·964	·186	4·00	·026	·08	·250	·0169
16	8	1·136	·180	−3·13	·032	−·14	·288	·0160

drawn for each element, and finally the values of dk_T and dk_Q for the various blade elements at any chosen value of J are plotted against the radial distance to obtain the thrust and torque grading curves along the blade. The integration of these curves gives the total thrust and torque of the airscrew, but a slight empirical correction is necessary to the thrust to allow for the drag of the airscrew boss.

16·22. Owing to the variation of the blade angle, chord and aerofoil section along the blade, it is not possible to obtain any simple analytical expressions for the thrust and torque of an airscrew, but the general nature of the airscrew characteristics can be examined by considering a typical blade element.

At zero rate of advance ($J = 0$) the axial interference factor a tends to infinity, since the axial velocity through the airscrew disc remains finite while the velocity V tends to zero. This occurs when

$$s\lambda_1 = 4 \sin^2 \phi,$$

and as ϕ is a small angle, this equation is approximately

$$sC_L = 4\phi^2,$$

where C_L is to be taken at an angle of incidence $(\theta - \phi)$. This state corresponds to a positive value of ϕ for an ordinary propulsive airscrew.

The other end of the working range of an airscrew occurs where the thrust vanishes at the point given by the equation

$$C_L = C_D \tan \phi,$$

so that the blade element is still carrying a small positive lift force. The torque is positive at this point, but vanishes at a slightly higher rate of advance, corresponding to the condition $C_L = - C_D \cot \phi.$

Between these two points the airscrew is acting as a brake, and beyond the point where the torque becomes negative, the airscrew acts as a windmill.

The efficiency of the blade element is

$$\eta = \frac{V.dT}{\Omega.dQ} = \frac{V}{r\Omega}\cdot\frac{\lambda_1}{\lambda_2} = \frac{1-a'}{1+a}\frac{\tan \phi}{\tan(\phi+\gamma)},$$

where $C_D = C_L \tan \gamma.$

This expression may be compared with the ideal efficiency $\dfrac{1}{1+a}$ deduced from the simple momentum theory of an airscrew and it appears that there are additional sources of loss of energy represented by

(1) a', the effect of the rotation of the slipstream,

(2) γ, the effect of the profile drag of the blades.

The first of these effects is small over the principal working range of an airscrew but the profile drag becomes of great

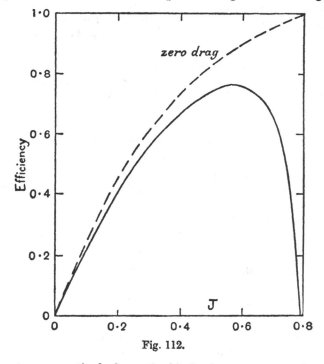

Fig. 112.

importance, particularly as the blade element approaches the attitude of no lift. Fig. 112 shows the efficiency of the blade element whose characteristics are given in table 21 and the broken curve represents the efficiency which would occur if the profile drag were zero.

The loss of efficiency can also be illustrated by considering the balance of energy for the blade element. The work done on the blade element in unit time is $\Omega\,dQ$ and this energy is distributed in the following parts:

$V.dT$, the useful work of the thrust,

$aV.dT$, the kinetic energy of the axial motion,

$a'\Omega.dQ$, the kinetic energy of the rotational motion,

dE, the loss of energy due to the drag of the blades.

The value of dE is obtained as

$$\begin{aligned}dE &= (1 - a')\,\Omega.dQ - (1 + a)\,V.dT\\&= \tfrac{1}{2}\rho W^2 Ncdr\,\{(1 - a')\,\Omega r\lambda_2 - (1 + a)\,V\lambda_1\}\\&= \tfrac{1}{2}\rho W^2 Ncdr.\,W\,(\lambda_2 \cos \phi - \lambda_1 \sin \phi)\\&= C_D.\tfrac{1}{2}\rho W^2 Ncdr.\,W,\end{aligned}$$

which is clearly the work done against the drag of the blade elements moving with the velocity W relative to the fluid.

The relationship between the circulation round the blade elements and the rotation of the slipstream is also of interest. The circulation round a blade element must be equal to $\tfrac{1}{2}C_L c W$ and the corresponding circulation of the slipstream is

$$K = \tfrac{1}{2}Nc W C_L = \pi s r W C_L,$$

while the total circulation of the slipstream is

$$\begin{aligned}K' &= 2\pi r.2\Omega a'r\\&= 2\pi r^2 \Omega\,(1 - a')\,s\lambda_2 \text{ cosec } 2\phi\\&= \pi s r W\,(C_L + C_D \cot \phi).\end{aligned}$$

Hence $$\frac{K'}{K} = \frac{C_L \sin \phi + C_D \cos \phi}{C_L \sin \phi}.$$

The circulation of the slipstream is due partly to the circulation round the blade elements which is associated with the lift force, and partly to the drag of the blade elements which tends to drag the air in the direction of motion of the blade. These two effects will be in the same ratio as the elements of torque contributed by the lift and by the drag respectively, i.e. in the ratio of $C_L \sin \phi$ to $C_D \cos \phi$, and this action of the drag of the blade elements accounts for the difference between the circulations K and K'. The vorticity of the airscrew disc is therefore of a complex nature, consisting partly of the circulation round the blades and partly of free vortex lines associated with the drag of the blades.

16·23. Calculations for a typical blade element can also be used to illustrate the variation of the characteristics of an airscrew with the experimental mean pitch. Fig. 113 gives

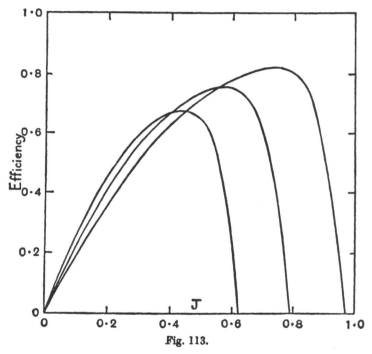

Fig. 113.

the efficiency curves for a blade element when the blade angle is increased or decreased 4° from its original value, and shows that an increase of pitch is accompanied by an increase of maximum efficiency. Aerodynamic considerations therefore indicate the advantage of using airscrews of large diameter and high pitch, but structural considerations limit the possibility of improvement in these directions. An airscrew is designed to absorb a definite torque at a definite rotational speed, and so an increase of diameter or blade angle must be accompanied by a corresponding decrease of the blade width. This process is clearly limited by the necessity for the airscrew blade to possess sufficient strength to resist the centrifugal and torsional stresses imposed on it, and the airscrew diameter is also limited by the fact that it is desirable to keep the tip velocity considerably below the velocity of sound. These difficulties can be avoided in part by the introduction of gearing between the engine and airscrew, so that the airscrew runs at a slower rate of revolution than the engine. The problem is, however, complicated by the weight and efficiency of the gearing, and a full discussion of the choice of the best airscrew in any given case is beyond the scope of the present treatise.

16·3. The aerodynamic theory has been developed for the case of a propulsive airscrew which gives a thrust in the direction of its axial motion, and it is necessary to examine whether the theory is also applicable to other working conditions of an airscrew. Fig. 114 shows diagrammatically the different types of flow which may occur with an ordinary propulsive airscrew at different positive and negative rates of advance. Type (2) represents the normal working condition, and as the axial velocity V increases the airscrew passes to the condition of type (1), where it acts first as a brake and then as a windmill. A different type of motion occurs when the airscrew has a negative rate of advance. Type (3) represents the flow for zero rate of advance which is a limiting case of the normal type (2), but as soon as the axial velocity V becomes negative a vortex ring will be formed round the airscrew as indicated by type (4) in the figure. For greater negative velocities the flow may correspond to type (5) or (6). The former represents the case when the airscrew gives rise to an eddy motion such as occurs behind a bluff body, and the latter represents a return to the initial type (1), but in the opposite direction.

The theory assumes the existence of a slipstream of conventional type and will be applicable to the types of motion (1) and (2). In the vortex ring state the momentum equations will break down both for the axial and for the rotational motion, and the thrust and torque of the airscrew will depend mainly on the rate of dissipation of energy in the vortex ring motion. The theory also breaks down for the motion represented by type (5) and will probably be only a rough approximation to the truth for type (3), which is the transition from the vortex ring state to the normal working condition. The final state (6) is similar to type (1) and the theory should be applicable to this case, but certain modifications are required to the momentum equations to allow for the fact that the direction of flow through the airscrew disc is reversed. In developing these equations the velocity u or $V(1 + a)$ represents the ve-

No.	Type of flow
1	
2	
3	
4	
5	
6	

Fig. 114.

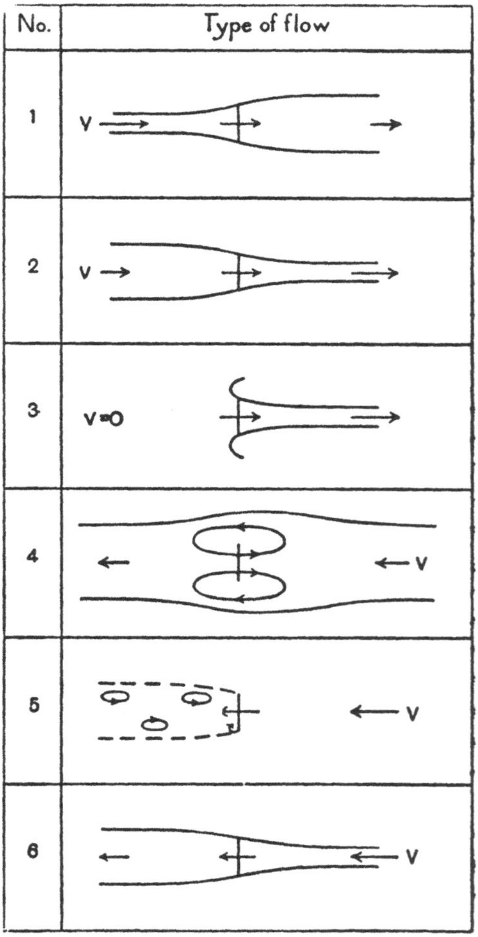

locity of flow through the airscrew disc and must be regarded as essentially positive. Hence for negative rates of advance the sign of the momentum expressions for the thrust and torque must be changed and this is equivalent to changing the sign of the expressions for $\dfrac{a}{1+a}$ and $\dfrac{a'}{1-a'}$ in 16·2. With this simple modification the theory may be applied to an airscrew with a negative rate of advance, provided that the conditions are such that a slipstream of conventional type is formed behind the airscrew. The condition for the validity of the theory is therefore that the value of the axial interference factor a shall satisfy the inequality $a > -\frac{1}{2}$. With this limitation the theory may be applied to any type of airscrew, irrespective of the purpose (propeller, windmill, fan, etc.) for which it is designed.

CHAPTER XVII

THE AIRSCREW: WIND TUNNEL INTERFERENCE

A model airscrew rotating in a wind tunnel disturbs the uniform flow produced by the tunnel fan and causes variations of velocity which extend to a considerable distance from the airscrew. This flow is constrained by the presence of the tunnel walls and the uniform axial velocity V which occurs at a sufficient distance in front of the airscrew in the tunnel differs from that which would occur in free air. It is necessary therefore to determine an *equivalent free airspeed* V', corresponding to the tunnel datum velocity V, at which the airscrew, rotating with the same angular velocity as in the tunnel, would produce the same thrust and torque*. A theoretical solution of this problem can be obtained by extending the simple momentum theory to the case of an airscrew rotating in a wind tunnel. The equivalent free airspeed is defined as that which gives the same axial velocity through the airscrew disc as occurs in the tunnel, since this condition will maintain the same working conditions for the airscrew blades, provided the interference effects of the rotational velocity are negligible. The equivalent free airspeed for an airscrew in a closed jet wind tunnel is normally less than the tunnel datum velocity†.

The assumption that there is no interference effect on the rotational velocity appears to be sound, but the representation of the interference effect by a change from the tunnel datum velocity to the equivalent free airspeed depends on the existence of the same axial velocity over the whole airscrew disc. This condition is satisfied approximately over the principal working part of the airscrew blades but fails towards the blade tips. In consequence the shape of the thrust

* Wood and Harris, "Some notes on the theory of an airscrew working in a wind channel," *RM*, 662, 1920.

† See Note 13 of Appendix.

and torque grading curves along the airscrew blade may be slightly different near the tip in free air and in a wind tunnel, although the method of correction is sufficiently accurate for the total thrust and torque of the airscrew. The method of correction would also appear to be unreliable near zero rate of advance of the airscrew, since the conventional type of slipstream assumed in the theory no longer occurs.

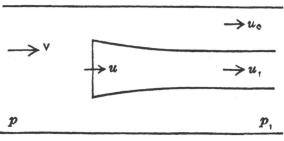

Fig. 115.

The type of flow assumed in the analysis is shown in fig. 115. V is the tunnel datum velocity, u the velocity through the airscrew disc, u_1 the slipstream velocity and u_0 the velocity in the tunnel outside the slipstream. The pressure rises from the original value p to the value p_1 in the region of the slipstream.

Let A be the airscrew disc area, S the cross sectional area of the slipstream and C that of the tunnel. Then by continuity

$$Su_1 = Au,$$
$$(C - S)\,u_0 = CV - Au,$$

and by use of Bernoulli's equation

$$\frac{T}{A} = (p_1 + \tfrac{1}{2}\rho u_1{}^2) - (p + \tfrac{1}{2}\rho V^2)$$
$$= (p_1 + \tfrac{1}{2}\rho u_1{}^2) - (p_1 + \tfrac{1}{2}\rho u_0{}^2)$$
$$= \tfrac{1}{2}\rho\,(u_1{}^2 - u_0{}^2).$$

Finally, the equation of axial momentum gives

$$T = S\rho u_1\,(u_1 - V) + (C - S)\,\rho u_0\,(u_0 - V) + C\,(p_1 - p)$$
$$= S\rho u\,(u_1 - V) + (C - S)\,\rho u_0\,(u_0 - V) + \tfrac{1}{2}C\rho\,(V^2 - u_0{}^2).$$

Now put $$\tau = \frac{T}{A\rho V^2} = \frac{4}{\pi}\,T_C,$$

and on eliminating u_1 and u_0 by means of the equations of continuity, the two expressions for the thrust become

$$2\tau S^2\,(C-S)^2\,V^2 = (C-S)^2\,A^2 u^2 - S^2\,(CV-Au)^2$$
$$= 2C\,(C-S)\,Au\,(Au-SV) - C^2\,(Au-SV)^2,$$

and

$$2\tau AS\,(C-S)^2\,V^2 = 2\,(C-S)^2\,Au\,(Au-SV)$$
$$- 2S\,(C-S)\,(CV-Au)\,(Au-SV)$$
$$+ CS\,\{(C-S)^2\,V^2 - (CV-Au)^2\}$$
$$= 2C\,(C-S)\,Au\,(Au-SV) - CS\,(Au-SV)^2,$$

from which it follows at once that

$$2\tau S\,(A-S)\,(C-S)\,V^2 = C\,(Au-SV)^2,$$
$$\tau S\,(CA-S^2)\,V^2 = CAu\,(Au-SV).$$

The equivalent free airspeed V' is such that it gives the same values to u and T. But in free air

$$T = 2A\rho u\,(u-V'),$$

or

$$(2u-V')^2 = \frac{2T}{\rho A} + V'^2$$
$$= 2\tau V^2 + V'^2.$$

Put

$$V = \lambda V',$$
$$x^2 = 1 + 2\tau\lambda^2,$$

obtaining for the free air condition

$$u = \frac{(x+1)\,V}{2\lambda}.$$

Also put

$$A = \alpha C,$$
$$S = \sigma A,$$

where α is generally small and σ lies between unity and 0·5, and the two wind tunnel equations become

$$4\,(x^2-1)\,\sigma\,(1-\sigma)\,(1-\alpha\sigma) = (x+1-2\sigma\lambda)^2,$$
$$2\,(x-1)\,\sigma\,(1-\alpha\sigma^2) = (x+1-2\sigma\lambda),$$

from which λ can be eliminated at once to give an equation for x in terms of α and σ, while the second equation then determines the value of λ.

The problem of determining the equivalent free airspeed has now been reduced to that of determining the value of λ for given values of a and τ by means of the subsidiary quantities x and σ. For this purpose the three equations are written in the form

$$\frac{x-1}{x+1} = \frac{(1-\sigma)(1-a\sigma)}{\sigma(1-a\sigma^2)^2},$$

$$\lambda = 1 + (x-1)a\sigma^2 - \frac{(2\sigma-1)x-1}{2\sigma},$$

$$\tau = \frac{(x+1)(x-1)}{2\lambda^2},$$

and a method of successive approximation may be used, trying different values of σ until the correct value of τ is obtained. As a guide to the value of σ, it may be noted that in free air σ would be determined by the equations

$$x^2 = 1 + 2\tau,$$

$$\sigma = \frac{x+1}{2x}.$$

In these equations a is the ratio of the airscrew disc area

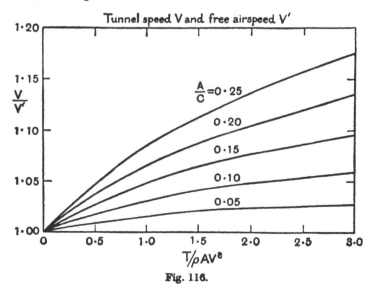

Fig. 116.

A to the cross sectional area C of the tunnel, and τ is the observed quantity $T/\rho A V^2$. The equivalent free airspeed is obtained finally by dividing the tunnel datum velocity V by the quantity λ. Curves of V/V' against $T/\rho A V^2$ for a range of values of A/C are given in fig. 116. The usual size of model airscrew tested in a wind tunnel corresponds to a value of A/C of 0·15 approximately.

The theoretical correction can be used for tests of model airscrews in a wind tunnel, but it is not possible to extend the theory to the case of an airscrew mounted in front of a body of any considerable size. Experimental work* has shown, however, that if the axial velocity is explored along radial lines just before and just behind the plane of rotation of the airscrew, the velocity tends to a limiting value which is equal to the equivalent free airspeed given by the theory when no body is present. This experimental method has therefore been adopted for the case of an airscrew-body combination and has been checked† by special tests in 4 ft. and 7 ft. tunnels.

* Fage, Lock, Bateman and Williams, "Experiments with a family of airscrews," part 2, *RM*, 830, 1922.

† Lock and Bateman, "The effect of wind tunnel interference on a combination of airscrew and tractor body," *RM*, 919, 1924.

APPENDIX

NOTE 1. (*See* p. 2.) It is now more usual to use the "quarter-chord point" as the point of reference for the measurement of moments. "The quarter-chord point" is the point on the chord line one quarter of the chord length from the leading edge.

NOTE 2. (*See* p. 39.) The contribution of the pressure and momentum integrals to the lift depends upon the shape of the large contour and the conclusion given on page 39 is not true for all shapes of contour; see Prandtl and Tietjens, *Applied Hydro- and Aeromechanics*, § 106.

NOTE 3. (*See* p. 95.) Since the publication of the first edition of this book a great deal of information on viscous flow and drag has been collected. This seems to show that vortex streets occupy a less significant place in the general picture than is indicated in Chap. VIII. For example, the wake of a circular cylinder takes the form of a vortex street in the range of Reynolds' numbers between 10^2 and 10^5, but at higher Reynolds' numbers the flow in the wake is turbulent but not periodic. Similarly, for aerofoils below the stalling incidence, a vortex street is only present in the wake for Reynolds' numbers below 10^5, which is outside the practical range. An account of modern work on this subject is given in *Modern Developments in Fluid Dynamics* (referred to elsewhere as *FD*).

NOTE 4. (*See* p. 108.) Blasius' empirical law for pipe flow is in good agreement with experimental results for Reynolds' numbers up to 10^5. It has however been superseded by a logarithmic resistance formula derived by Karman which has a sounder theoretical basis and gives results in better agreement with experiment for Reynolds' numbers above 10^5; see *FD*, § 154.

NOTE 5. (*See* p. 115.) This formula for the drag of a flat plate has been superseded by the formula

$$C_D = 0.91 \left[\log_{10} \frac{cV}{\nu} \right]^{-2.58}$$

which is based on a logarithmic resistance formula derived by Karman; see *FD*, § 163.

NOTE 6. (*See* p. 116.) It is implied in the paragraph in the text that a vortex street is always formed behind a bluff body, but it is now known that this is not the case. (*See* Note 3.)

NOTE 7. (*See* p. 118.) Actually there are small changes of pressure across the boundary layer on a curved surface; see *FD*, § 45. Also, due to the presence of the boundary layer, the pressure at the trailing edge of an aerofoil with a finite trailing edge angle does not rise to the full theoretical stagnation pressure.

NOTE 8. (*See* p. 119.) For laminar boundary layers and specified pressure distribution the position of the separation point is independent of the Reynolds' number; see *FD*, § 49.

NOTE 9. (*See* p. 120.) For aerofoils at small incidences the pressure on both upper and lower surfaces increases towards the trailing edge; this does not cause a breakaway, which only develops on the upper surface at high incidence.

NOTE 10. (*See* p. 122.) The motion of an aerofoil starting from rest is well illustrated by Plates 17–22 of *Applied Hydro- and Aeromechanics* by Prandtl and Tietjens.

NOTE 11. (*See* p. 124.) The drag of an aerofoil depends on the extent of the laminar boundary layer, and the drag coefficient can only be less than the frictional drag coefficient of a flat plate at the same Reynolds' number if the laminar boundary region is more extensive on the aerofoil than on the flat plate.

NOTE 12. (*See* p. 189.) There are now wind tunnels with both open and closed working sections at most of the larger aeronautical research establishments.

NOTE 13. (*See* p. 222.) For an airscrew working in an open jet wind tunnel it is usual to assume that the tunnel interference effect is negligible. The absence of any correction to the equivalent free air speed is an advantage and for this reason airscrew tests are now generally made in open jet wind tunnels.

NOTE. (*See* p. 88.) The vorticity which occurs for the straight line aerofoil is assumed to be proportional to cot $\frac{1}{2}\theta$ in the text without any proof being given. The validity of this expression is verified by the result that it produces a constant induced velocity along the chord of the aerofoil. A direct proof of the result is given in *Aerodynamic Theory* (edited by Durand), Vol. II, Division E, pp. 37–39.

BIBLIOGRAPHY

Some of the better known modern books on low-speed aerodynamics are listed below.

DURAND (ed.).	*Aerodynamic Theory* (6 vols.), 1943.
GOLDSTEIN (ed.).	*Modern Developments in Fluid Dynamics* (2 vols.), 1938.
LAMB.	*Hydrodynamics*, 1932.
MILNE-THOMSON.	*Theoretical Hydrodynamics*, 1949.
MILNE-THOMSON.	*Theoretical Aerodynamics*, 1948.
PIERCY.	*Aerodynamics*, 1947.
PRANDTL.	*The Essentials of Fluid Dynamics*, 1952.
PRANDTL & TIETJENS.	*Fundamentals of Hydro- and Aeromechanics*, 1934.
PRANDTL & TIETJENS.	*Applied Hydro- and Aeromechanics*, 1934.
ROBINSON & LAURMANN.	*Wing Theory*, 1956.
SCHLICHTING.	*Boundary Layer Theory*, 1955.
THWAITES (ed.).	*Incompressible Aerodynamics*, 1959.

INDEX

Moment at zero lift, 86, 91
„   coefficient, 2
„   , general formula for, 80
„   of an aerofoil, 2, 82, 91, 150, 179
Monoplane aerofoils, 137
Motion, equations of, 109

Outflow, 209

Pipes, flow in, 107
Pitch, experimental mean, 199, 218
Potential function, 54
„   , velocity, 48, 55
Pressure, centre of, 2
„   , dynamic, 2
„   , total head, 11, 42
Profile drag, 7, 123, 141
Propeller, 200

Reynolds' number, 105

Scale effect, 105, 120, 123
Singular points, 61
Sink, 21
Slip on boundary, 100, 117
Slipstream, airscrew, 201, 208
Slug, 9
Solidity of blade elements, 213
Sound, velocity of, 12
Source, 21, 49, 55
„   and sink, 27, 51
Speed, equivalent free air, 222
„   , measurement of, 11, 15
Stagger, equivalence theorem for, 181
Stagnation point, 14

Steady motion, 10
Straight line and circle, 64
„   „   , flow past, 66, 94, 119
Stream function, 18, 49, 55
„   line, 10, 50
„   tube, 10, 15

Tailsetting, angle of, 196
Thrust and torque coefficients, 199
Transformation, conformal, 58
„   , Joukowski's, 71, 77
„   of flow pattern, 63
Turbulent flow, 108, 115
Two-dimensional motion, 18

Uniform loading, 134, 157
Units, 9

Velocity field, 45
„   , induced, 45, 127, 132, 156
„   , potential, 48, 55
Viscosity, 5, 99
„   , coefficient of, 100, 103
Vortex, bound, 129
„   , induced velocity of, 45, 127
„   , line, 126
„   , point, 44, 50, 55
„   sheet, 47, 117
„   street, 95, 116, 119
„   strength, 39, 126
„   , trailing, 129
„   tube, 125
Vorticity, 39, 125
„   , constancy of, 41

Windmill, 200, 206
Wind tunnel interference, 189, 222